Marxism and the Individual

Cem Eroğul

Marxism and the Individual

Translated by Paula Darwish

palgrave
macmillan

Cem Eroğul
Istanbul, Turkey

Translated by
Paula Darwish
Manchester, UK

ISBN 978-3-030-83661-0 ISBN 978-3-030-83662-7 (eBook)
https://doi.org/10.1007/978-3-030-83662-7

This Palgrave Macmillan imprint is published by the registered company Springer Nature Switzerland AG.
The registered company address is: Gewerbestrasse 11, 6330 Cham, Switzerland

To all those who fight for the creation of a social order where "the free development of each is the condition for the free development of all."
(The Communist Manifesto)

Preface

Just a few words to introduce myself. I was born in 1944, in İzmir-Turkey. As I turned fifteen, I realised that I was most dissatisfied with myself, physically and intellectually. Deciding to thoroughly reshape myself, I bought a book about physical exercise. It turned out to be an excellent book,[1] dividing the body into twelve muscle groups and explaining thoroughly how to develop each of them. I started immediately and since then I have always kept a watchful eye on my body. When it came to the intellectual reformation, I reasoned that to be able to educate myself I would have to start with the science of education, i.e. pedagogy. But I soon realised that I was going down a blind alley. I then turned to psychology. After a while, I came to see that I would not get very far while so centred on myself. What I needed was a general conception about the world, including myself. I needed a worldview, or in other words a philosophical outlook. This took me to philosophy. I read extensively. All this reading was done in French. Being fluent in French provided me with a privileged access to the main sources of Western culture. Nevertheless, when I graduated from the Faculty of Political Sciences in Ankara, the sad truth was that despite all this reading, I had reached my twentieth birthday almost as confused as I was at fifteen.

In 1964, I went to Paris owing to a scholarship granted by the French government. I enrolled for a DES (diplôme d'études supérieures) at the Faculté de Droit et des Sciences Économiques. By the time I had completed my MA at the end of 1965, I had discovered Marxism and had at last found the fundamental conception of the world that I had been so eagerly seeking since the age of fifteen. From then on Marxism has oriented all my life.

In 1966, I started an academic career at the Faculty of Political Sciences, in Ankara. I had just become a full professor in 1983, when I was kicked out of the university during the fascist regime established by the military coup of 1980. I was reintegrated in 1990 with the return to a relative democracy. I retired in 2010. From 1966 to 1971, I was an activist in the Turkish Labour Party. I was cut off from active politics after the party was banned in 1971. From then on, my Marxism was mainly limited to the intellectual field. All my teaching, all my writing, has always been inspired by Marxism. In 1969, I produced the first monograph of a political party ever written in Turkey. It was a class and ideological analysis of the so-called Democrat Party, which governed Turkey between 1950 and 1960. It was a Marxist study which did not declare itself openly as I was prosecuted for communist propaganda in those years. In the early seventies, I started systematic research aiming at the elaboration of a Marxist theory of politics that would satisfy me. This endeavour resulted in the publication of *An Essay on the Nature of the State*,[2] a book written in English at the end of an extended sojourn in England between 1979 and 1981. With this book, I came to the conclusion that from now on the very existence of the state itself was the main problem, rendering thus futile any search for more democratic forms of state rule. The spread of capitalist globalisation, the demise of the so-called socialist states, and the emergence of totally new forms of contestation resulted in my gradual dropping of politics as a main theoretical object. With traditional politics as we knew it in a state of collapse, after more than five thousand years of state rule, it dawned on me that the "social individual"[3] was becoming the decisive agent in the establishment of the future post-capitalist society.

This meant that from now on, the individual as such was to be the main object of Marxist theoretical research. Starting from 2002, I endeavoured to build a theoretical framework which would serve this purpose. During the first years, I was totally submerged and nearly drowned in reading in so many directions: in anthropology, psychology, and Marxist classics. It was only in 2007 that I was able to construct the main frame of my approach. I presented it to the Tenth Conference on Social Sciences, held that year in Ankara. This paper of thirty or so pages was entitled "The Individual as a Form of Existence of Humankind".

But I was still far away from a theoretical draft that would enable me to cover the main aspects of the problem. It took me another seven years to develop it. I finally reached a general theoretical grasp of the concept of the human individual according to Marxism. It was completed in the main but it still needed some elaboration. But how long could I pursue a seemingly endless endeavour at my age? Those who followed my research advised me to round up my book as soon as possible. Eventually, I thought it wise to follow this advice. That is how the Turkish original of the present book was published in 2014, as a kind of jubilee for my seventieth anniversary. The Turkish version comprises some appendices aimed at the Turkish reader. I dropped them for the English version. What remained was a text of 320 or so pages. I first reckoned that it would be more convenient to rewrite the text for the English reader, in order to present my essential argument in a condensed form. But this was not to be. I had to stop at the end of the first chapter. Old age prevented me from completing the present endeavour. So I hastily reviewed the Turkish text and prepared an eventual second edition, leaving the English version to the care of a translator who will be able to use this last revised text.

Istanbul, Turkey Cem Eroğul[4]

Notes

1. Marcel Rouet, *Santé et Beauté Plastique* (Paris: Editions J. Oliven, 1957).
2. Accessible on the web: cemerogul.wordpress.com
3. "... it is, in a word, the development of the social individual which appears as the great foundation-stone of production and of wealth." Marx (1973, p. 705).
4. To be pronounced "gem" (skip the meaning!) "errohool".

Acknowledgements

Thanks to Hayri Erdoğan and Gökhan Atılgan for offering their absolute support from the very beginning of a project which took some twelve years to complete, and for publishing it with Yordam Kitap on my seventieth birthday. I am grateful to Gönül Göner from Yordam Kitap for diligently providing technical support for the restructuring of this book's hastily published first edition into a second edition, on which this English translation is based. Thank you also to Gökbörü Sarp Tanyıldız for believing in my work and referring it to Terrell Carver. I am indebted to Terrell Carver for appreciating my endeavour and bringing this book to the attention of Palgrave Macmillan. This book would not be possible without the editorial support of Philip Getz and the staff of Palgrave Macmillan. Finally, thanks to Paula Darwish for her translation and strenuous efforts to remain faithful to my intended meaning.

Contents

Introduction

The standard definition of the human as a "bio-psycho-social being" is in fact not a definition. It is just a description. Everyone has always observed that as a species we are endowed with a specific body, we can feel and think, and we cannot survive without being cared for by others. But such a description does not tell us what is our most important part, the part that makes us humans. As long as we do not know this, we cannot grasp our peculiarity nor what specifies us as a species. Aristotle, with his usual thoroughness, sensed the problem and defined the human as a "zoon politikon". Thus, for him, what makes us humans is that we live as part of political aggregates. But he was mainly misled in the generic part of his definition. By calling our species a "zoon", he got entangled in a hopeless contradiction: how could an "animal" ever be political? On the other hand, as Marx and Engels wrote in *The German Ideology* (1846), "Men can be distinguished from animals by consciousness, by religion or anything else you like." (Politics for that matter.) But, "They themselves begin to distinguish themselves from animals as soon as they begin to produce their means of subsistence."[1]

The definition of humankind could only be derived from *this fact* established by evolution itself. But one must be very careful in "reading" this fact. A lion obtains its "means of subsistence" by chasing its prey whereas humans "produce" it. Contrary to the single lion hunting its prey, such a production implies the conscious use of the means of production by a group of individuals. Production can only be "social" production, and to

C. Eroğul, *Marxism and the Individual*,
https://doi.org/10.1007/978-3-030-83662-7_1

be social definitely implies the use of a mode of communication. Thus, there cannot be production without the existence of some kind of language. Language that enables intercourse in the process of production is also the source of individual consciousness. Right from the outset, the social can only exist in the form of distinct conscious individuals. Production aims at the satisfaction of needs, but human intercourse constantly creates new needs. The satisfaction of needs incessantly reshapes the individual, both body and soul. What defines the human body is not what it has in common with animals, but the specific form imparted to it by social intercourse. What is commonly called the "soul" is the individual form, shaped and constantly reshaped, through social relations. Therefore, in all its aspects the individual is the product of social intercourse. In that case, it becomes clear that humankind is the only existing "social" species. It is unique. It is not a kind of animal. Trying to define humans as animals endowed with specific properties is as ridiculous as defining animals as "moving plants".

This Marxist definition of the human presents a real challenge to any scientific inquiry concerning anything human. As the social is always the individual, and as the individual is always the social, any approach used must be able to grasp these two aspects of the human simultaneously. But at the same time, allowance must be made for grasping the specificities of these two aspects separately. Following a lengthy search, I finally found the solution of this riddle in Marx himself. After much effort, it dawned on me that the method of analysis used by Marx in *The Grundrisse*,[2] in order to sort out the identity and the specificities of production and consumption, is exactly what is needed in the approach to the human reality of both the social and the individual. But Marx nowhere explicated the methodological frame of this method of analysis. He was just content to use it wherever it appeared convenient to him, as for instance was the case in the analysis of relations between commodities and money.

Thus, the first requirement in the transposition of this approach into another field of research was to strip it from its concrete peculiarities and posit it as an abstract methodological model. I undertook this task in the first chapter of the book, which is devoted to working out the main cognitive tools necessary for my analysis of the Marxist concept of the human individual. This tool can be labelled "dialectical identity", and I attempted to reformulate it in the context of a theory of the "social individual". This is the subject of the first section. But a few other tools were needed as well. Their task would be: on the one hand, to help me figure out the modalities of the incessant transmissions between the individual and the social; and on the other hand, to give me specific conceptual tools for my analysis of human social and individual aspects when considered on their own. Fortunately, I had already elaborated these tools during my research on the Marxist concept of politics and had explained and used them in my *Essay on the Nature of the State* (1981). All these tools are presented and briefly explained in the following sections of the first chapter.

The second section of the first chapter is devoted to the problem of finding regularities in the uninterrupted flow of individual/social activities. I found it useful to employ some of the most widespread concepts of social sciences, such as roles, structures, and institutions, taking care to redefine them within a Marxist perspective. In the individual, these concepts, which indicate more or less durable social crystallisations, take the form of habits. They allow us to conceptualise the incessant flow of human activity as a process of coagulations and dilutions. In the third section, I present a thinking tool that I found extremely useful in analysing determinations. Be it on the social or on the individual level, interminable discussions have often plagued Marxist thinking about determinations. I found that the most handy distinction is between active and passive determinations. The active determination accounts for the existence of a phenomenon, and the passive one conditions it, thereby shaping its specific nature. In this section, I tried to provide some refinements in the construct of this twin-concept in the hope of clarifying its exceptional usefulness in practical analyses. The fourth section deals with the relations between quantity and quality. The dialectical conception of qualitative transformation stemming from quantitative increase or decrease is a precious source of awareness which is often overlooked. It is not only helpful in the clarification of developmental processes, but it also helps to understand, among other things, how a passive determination can be metamorphosed into an active one. The fifth section deals with the abstract/concrete dialectics in the conduct of theoretical analyses. Many errors and misunderstandings in Marxist thinking, and indeed practice, have their source in the mishandling of this problematic. Any analytical tool should be carefully checked before its application, questioning in particular its relevance to the level of abstractness/concreteness under consideration. The same goes for very practical matters as well. To give a couple of vulgar examples, being against private property does not legitimise theft, and opposition to the family as an institution is not a licence to cheat on one's spouse, unless different levels of analysis were to be mingled. In the sixth and last section of this chapter, I round up the many discussions concerning diverse cognitive devices by stressing that "sociality" is indeed the defining trait of the human species. My aim in this chapter is to reach a sufficient degree of clarity concerning the cognitive conceptual framework used in the following two chapters for analysis of the human being, in both its individual and social aspects.

The second and third chapters respectively are devoted to the analysis of the individual and social aspects of humankind. Consequently, the second is entitled "the soul" (or the psyche), as this is the peculiar form taken by the human individual, whereas the third chapter is entitled "culture", taking into consideration that the human soul is always shaped through culture. That said, I must immediately warn the reader against this apparent separation between the two aspects of humanity that are in fact inseparable. To palliate any such misunderstanding of division, which may arise from my mode of presentation, I frequently deal with culture in the second chapter, just as I also frequently refer to the psyche in the third chapter. The body does not warrant a place in the main titles. The reason should be easy to understand as it does not constitute a third aspect, being in fact the exclusive bearer and representative of the soul, as well as being the main, but not exclusive, bearer and representative of culture. Thus, we will quite frequently discuss the body, both in the second and third chapters.

Another seeming inconsistency is the internal division of the second and third chapters into two main parts, entitled respectively "theory" and "literature", as if any theoretical view could be elaborated without making use of the literature of the field. Even so, I opted for such a presentation with a very pressing consideration in mind. The fact is that the literature about the psyche and culture is indeed immense. I could not eschew the obligation to pick the handful of sources that were particularly relevant to my approach to the psyche, as well as to my use of culture for building a theory of the human individual. Abstaining from such a selection would leave the reader drowning in a sea of disparate sources, serving thus neither clarity nor fairness. Moreover, making these selections has given me the possibility to assess each selected source with a twin concern in mind. Criticising the aspects not aligned with my point of view permitted me to further clarify my approach, while the use of parts I approved of allowed me to further develop the Marxist conception of the individual as I have elaborated it.

The theoretical parts of the second and third chapters are divided into ten and nine sections respectively, each stressing different dimensions of the object under analysis. But as the object of my study is in fact one and the same, namely, the human species both in its individual and social aspects, a lot of repetition has been not only inescapable but in fact quite necessary. The literature parts of the second and third chapters take respectively eight (the third subtitle covering in fact six different approaches) and seven main contributions into consideration. Each of these contributions is generally due to a single author, but I have sometimes found it useful to call on some additional sources.

The object of this research is truly immense. I covered as much as I could. I have presented in the conclusion a brief summary of what I consider to be my contribution. I have also indicated some paths which appear to be promising for further developing this research. It is my dearest hope that this study will offer even a modest contribution to our common task of building a truly humane society.

Notes

1. Marx and Engels (1998, p. 37).
2. "[*Consumption and Production*]". Marx (1973, pp. 90–94).

References

Eroğul, C. (1981). *An Essay on the Nature of the State*. University of Ankara, Faculty of Political Sciences Publications.

Marx, K. (1973). *Grundrisse*. Random House, Inc.

Marx, K., & Engels, F. (1998). *The German Ideology*. Prometheus Books.

Chapter I: Cognitive Tools

Section 1. Dialectical Identity. Identity Via Specificity

As indicated in the introduction, the standard definition of the human as a "bio-psycho-social being" is most unsatisfactory. It is not a definition, it is just an observation. My inquiries into the Marxist understanding of the human started exactly with the frustration I felt in the face of such a statement. Of course, Marxism is far from denying the many-sidedness of objects. But all the diverse properties must be accounted for by a *dominant property*. Otherwise, rather than being confronted with a theoretical explanation, we are left with a hollow eclecticism. Let's take as an example the key concept of *Capital*: the commodity. As we all know, the commodity possesses two forms of value: a use value and an exchange value. A commodity deprived of either of these values is no more a commodity. Thus they are both indispensable. But which one truly *defines* a commodity? No object without use value could ever be a commodity. But all objects having a use value are far from being commodities. It is only when a useful object obtains an exchange value that it becomes a commodity. The same goes for humans. Evidently, without a body the human species could not exist. But what makes it human is not its body; it is its social life. Not only an aggregate, but a "social" life; i.e. a community constituted by differentiated individuals cooperating among themselves. Just as we cannot grasp the nature of commodities by inquiring about their use value, neither can

C. Eroğul, *Marxism and the Individual*,
https://doi.org/10.1007/978-3-030-83662-7_2

we reach a true understanding of the human by studying its bodily properties. Throughout this book, we are going to analyse the biological, psychological, and social aspects of human reality. But to eschew any eclecticism, it must be stated right from the outset that what defines the human, what explains not only its social aspects but its biological and psychological aspects as well, is the fact that it is the only *social* being existing on earth. Humankind owes its defining particularity to the fact that it is the only species to create itself by taking the path of fulfilling the needs it shares with other animals through social production of the means to satisfy them. It is in this process that humankind took the form of a community of distinct conscious individuals cooperating in order to satisfy needs ever renewed and expanded as a result of this cooperation. It follows that for a theoretical research endeavouring to grasp human reality, the main task is to devise an approach capable of embracing both the social and the individual at once. Inquiries focusing exclusively on the social might not be under such a binding imperative, at least in the beginning. But when the focus is on the individual, an approach which does not take into account the social nature of the individual right from the first steps of the most abstract level is doomed to fail.

Whenever confronted by a similar task, Marx solved the problem by making use of what can be labelled the "dialectical identity" approach. As cited in the introduction, a brilliant instance of such a usage is to be found in the *Grundrisse* where Marx analyses the correspondence between production and consumption. His analysis is conducted on three levels. At the first level, we are faced with a direct identicality. Production is consumption and consumption is production. He gives the example of a person feeding herself. Eating is a consumption which replenishes bodily reserves, thus realising the reproduction of the body. At this level, one and the same action is simultaneously both production and consumption. At the second level of analysis, production and consumption act as a means for each other. Production furnishes the object to be consumed. Consumption determines what object is to be produced. For instance, the cultivator grows plants which serve as food-stuffs. But it is the consumer who determines which plants are to be cultivated as food-stuffs. At the third level, a total reciprocal determination occurs. Production and consumption are not only identical, not only the instrument of each other, but they achieve their true realisation by going through a process of mediations which culminates in the transformation of one into the other. Production realises itself as true production only when the product is finally consumed, and it is this final consumption which determines the quality and quantity of what is to be produced next.

It is crucial to grasp clearly the vigour of this approach. This is not a play on words, but a truly powerful cognitive device. A trend of thought that starts from identity, and ends up with the reciprocal determination of the terms of that identity, allows the developments of each term to be analysed separately, while constantly keeping their fundamental unity in mind. Pursuing the example just given, it is only by keeping in mind the identity of production and consumption that one can wonder why it is that they have to go through the mediations of exchange, distribution, and circulation to be able to realise their identity. (Whereas this identity is almost immediately realised in the example of the primitive communal mode of production.) It is this questioning which induces us to trace the role of the mediations, and it is this research which opens the gates of scientific discovery. Thus, Marx uncovered the mechanisms of the capitalist mode of production by analysing the functioning of the mediations it imposed on the process that occurs from production to consumption and vice versa. This method, this dialectical identity approach, appears to be most promising whenever the object under study is a twin reality. A couple of examples immediately come to mind: the pairing of philosophy/sciences is such an example. Philosophy can only be defined as a compendium of all current knowledge, i.e. the sum total of sciences. This is the first level. But to become truly knowledgeable, philosophy and each science must be studied separately. This is the second level. However, true philosophy must encompass all the leading sciences in their mature form, just as on the other hand each science needs a broad philosophical overview to achieve an authentic development. How this goal is pursued, for example in the universities of a definite country in a definite historical period, constitutes the third level of the dialectical identity analysis. Another subject matter could be the gender/person twin reality. At the first level, the fact that every person is necessarily gendered has to be described as a given fact. At the second level, the different meanings of gender (natural, assigned, claimed) and their varieties are to be studied in the context of their instrumentality for the person and vice versa. Finally, at the third level, the object of analysis will be to discover through what kind of mediations specific genders and specific persons come to get stuck with each other in a concrete social setting and what this results in. As can be easily seen, the dialectical identity approach can be tremendously productive when it is applied to an adequate object and when the general warning about congruence in the level of generality-abstractness and particularity-concreteness is carefully heeded.

In order to be able to use the dialectical identity approach of the *Grundrisse* in other contexts, it is first necessary to expose it abstractly. I think that Marx's method can be accurately outlined as follows: in dialectical identity, the analysis must be conducted on three different levels, each level taking into consideration a different kind of identity. The first level concerns itself with "direct identity". For example, every way up is a way down (depending on the direction you face) or life is death (as cells renew themselves while sustaining life) and death is life (as decomposition of the dead body realises itself through the bourgeoning of new forms of life). The second level has to do with "instrumental identity". One must climb first to be able to descend and vice versa, or life cannot go on without an alternation of births and deaths. The third level deals with "mediated identity"; here, identity cannot be simply stated or its reciprocal instrumentality generally indicated. The third level requires a concrete elaboration explicating the concrete paths taken by each term of the identity in order to achieve its identification with the other term: What kind of slope? What kind of movement? Who or what goes up or down? Velocities, accelerations, etc.; or Who or what is born? Who or what dies? Under what conditions? Periodicity, rhythms, etc.

At first sight, this three-tier type of analysis can look quite simple. But it is not so, because the elements of each tier and their modes of interaction vary according to the object under study, and this knowledge has to be fetched up from the particular science dealing with this object. Dialectics can be useful only if it leans on firm scientific knowledge. Illustrations will not suffice. And analogy can be quite misleading at this level. Rather than analogy, thorough analysis focused on the subject matter is an absolute must. Only then can every tier of the dialectical identity model be constructed with adequate material. This is the first trap to watch out for if one does not want to finish up down a dead-end.

There is also a second trap that must be carefully watched out for when using this method; it is a general danger threatening all kinds of scientific analysis and looms here as well. It consists in the possible confusion between levels of abstractness/concreteness. It is most important to be consistent when dwelling on a specific layer of abstractness or concreteness. Jumping, or involuntary slips, from one layer to the other can easily lead to error. Again in the *Grundrisse*,[1] Marx masterfully exposes how to tackle the passage from one level to the other both in research and in the mode of exposition. A statement true at one level can be wrong at another. Let us imagine as an example that we want to analyse a specific political party of a particular country in a given historical period. We may well start with a general statement concerning political organisations. But we cannot jump from this general statement to the particular party under study. We must first check if this statement holds for political parties as a specific mode of political organisation. Yet, this first step towards more concreteness may well be insufficient. This second statement as well must be translated into the language required by the level of concreteness we are dealing with. It is exactly this imperative that Lenin has in mind when he stresses the necessity of conducting "concrete analyses". Otherwise, one can easily skip over some determinations which can prove to be crucial, both in the analysis of a situation and the choice of the strategy to be adopted for dealing adequately with it. As noted in the introduction, and as will be insisted upon in the fifth section of the present chapter, keeping a keen eye on the level of abstractness-generality or concreteness-particularity in analyses is of crucial importance both in scientific analysis and in praxis.

I kept a watchful eye on these two caveats in my attempt to adapt the three-tier dialectical identity approach to the study of the individual/social twin aspects of the human species. The general conclusions I reached are as follows:

According to the *first level,* the individual is the social and the social is the individual. There is a "direct identity" between them. No-one has ever witnessed a society without individuals and individuals can never become human, indeed neither can they even survive, unless they are in relations with other individuals. In short, there is no society without individuals, and no individual without society. It seems to me that the main lesson to be derived from this basic truth is *negative*. This constitutes a "warning"; it tells us that any theories of the individual which overlook society, or any social theories which do not consider the individual, are basically flawed and must be discarded from the outset. This is particularly true for all varieties of theoretical constructions relying on some Robinson living in a secluded island as a model of the human species. Since an individual not shaped by social intercourse cannot exist, all approaches starting with a "human individual" who is not *already* socially determined by their social environment are just unscientific. This is equally true for all kinds of social theories that imagine they can elaborate on the human without taking into consideration the individual. The source of social energy is always individual. Social action and change can only be carried out by individuals.

Thus, a social theory ignoring the individual will inescapably end up with a petrified object, or it will have to invent supra-individual entities such as "society", "class", etc., allegedly acting in lieu of concrete individuals. As seen, the first tier of the dialectical identity analysis functions mainly as a salutary warning lighthouse, imparting on us the necessity of keeping well away from any theoretical claim starting from a non-already-social individual, or a non-individually mediated social whole.

The *second level* is the level of "instrumental identity". The positive building up of our cognitive device starts at this level. The challenge is to clarify the meaning of this double assertion: "The instrument of the social is the individual" and "the instrument of the individual is the social". Put differently, "the social can only constitute itself through individual action" and "the individual can only become an individual through social intercourse". The only way to clarify these assertions is to elaborate a scientific construction using mainly the theory of evolution and the teachings of diverse sciences such as archaeology, palaeontology, geology, climatology, genetics, microbiology, etc. All these sciences supply us with various data that help to elucidate the problem of how the human species emerges and develops. An overview of the emergence and evolution of the human species can be summarised as follows:

Around some five to six million years ago, the hominids branched out of their common ancestor, which they shared with chimpanzees and bonobos. What distinguished the hominids was their bipedalism. This new way of locomotion was accompanied by bodily transformations, and it became firmly established within a period covering roughly two and a half million years. Around two and a half million years ago, the hominids intensified their use of differently shaped stones for hunting, digging, uprooting, crushing, squashing, etc., starting to reshape these stones to make them more suitable for various tasks. The new stone instruments were mainly produced by purposefully striking stones against rocks with the aim of breaking them into suitable forms. This is a dramatic turning point, as it inaugurates a purposeful fabrication of tools by an animal species, something unseen up to this time. What rendered this development possible was of course the radical adoption of bipedalism, permitting the complete liberation and evolution of hands. The definite establishment of stone-tool manufacture occurred around one million nine hundred thousand years ago. This is indeed the turning point taking us from the hominids to the homos. The hominids do not disappear at once, but they start their journey towards total extinction. From then on, the evolution of the homo species is governed by the requirements of a mode of subsistence relying on the purposeful fabrication, and use, of means of production realised through cooperation within a definite group of human individuals. As pointed out in the introduction, such a cooperation implies the emergence of a mode of communication, a language. Language prepares the ground for the capacity of symbolic thinking to develop. The human species uses language to share its knowledge and to transmit it to the new generations. Thus, gradually cultural evolution takes precedence over biological evolution. It is crucially important to be absolutely clear about this mechanism. Evidently, social life does not drive biological evolution. Mutations are always random. But natural selection decides which mutations will thrive and which will not survive. A determining factor when it comes to the human species is the fact that "natural selection" is commanded by the requirements of social life. Changes boosting social life are encouraged and thus thrive, whereas impediments to sociality are left to die out. What is more, with the continuous acceleration of cultural progress, the role of biological evolution tends to lag dramatically. Humankind takes over its own evolution from the realm of biology and starts to shape it more and more consciously. What this brief summary makes clear is that without social life, without a cooperative and purposeful production of the means for satisfying needs (which were first shared with other animals but progressively developed by social intercourse), the human species could not have come into existence. And as purposeful production and social cooperation inescapably necessitate the existence of distinct conscious individuals communicating through language, this social life could have only one form: a cultural network bringing together distinct individuals.

The emergence of humankind as a particular species produced an incredible impact on the world. The fact that humankind cannot exist without modifying their environment as well as themselves brought on tremendous transformations all over the world. All plants cultivated today, regardless of type and location, have originated from seeds consciously altered by our ancestors; even those that grow naturally are already profoundly marked by human intervention, be it the result of ignited fires, induced floods, contaminated water, poisoned air, or altered climatic conditions. The so-called "natural flora" is by and large a human product. The current much-trumpeted problem of genetically modified organisms is, in fact, merely the actual form taken by a mode of intervention which goes back thousands of years. The same applies even to a larger extent to the animal world. There is not much left from the so-called "natural fauna".

Apart from insects, worms, birds, etc., the vast majority of animal species making up our environment is made up of animals thoroughly altered by human intervention. Again, interventions like cloning, which until recently were considered as sensational news, are in fact merely one of the last avatars of a human activity stretching over millennia. Animals living in the wild have almost totally disappeared. Those trying to survive in zoos rapidly lose their original characteristics. They are subdued, tamed, and degraded. They inevitably degenerate. Nature, with all its plants and animals, has become more and more part and parcel of our *social environment*. The same goes for our inorganic environment as well. The industrial age has thoroughly reshaped our landscape, with urban life invading the earth, and pollution threatening the soil, water, and even the air. The human body itself was one of the main objects of this reshaping of nature. And this is also the result of human agency. Social production, the new forms taken by foraging and hunting, and the satisfaction of ever-renewed needs are the main factors explaining the evolution of hands, larynx, and brains. Mutations favouring this evolution were retained, whereas others thwarting it were left to die out. The human genome is identical to the genome of the chimpanzee to a degree well exceeding 95%. This is amazing when the phenotypes and the way of life of these two species are compared. It is almost unbelievable that such great disparity could be accounted for by such a small proportional difference. This is striking proof of the fact that it is not the body that defines humans. If it were so, humans' and chimpanzees' genomes would not have been over 95% alike. Moreover, as to be expected, not only the generic body of the human species but all individual bodies are also social products. Within the generic boundaries traced by the evolution of the species, every individual body is the product of *socially regulated* mating. Gestation and giving birth are also socially regulated. But genuine humanisation occurs after birth. All the conditioning, shaping of perceptions, implanting of habits and capabilities, in particular upright walking and speech capacity, and in other words, every characteristic of the species is inculcated into the individual through social intercourse. What is true for the body is even truer for the soul or the psyche. Our subconscious, our emotions, our feelings, our conscious life are all social products. The fundamental truth is that as part of its own evolution, the human species has incessantly transformed nature, its own body and psyche, as well as its social life. The current general collapse threatening our natural and social environments and the sense of global madness enveloping us are only witness to the incredible acceleration attained by our transformative powers.

Another requirement of the second level of dialectical identity analysis is to determine the particular properties of each aspect of the identity. In philosophical language, this constitutes the "metaphysical" phase of dialectical analysis. As is well known, according to Hegel the proper metaphysical method is to study objects on their own, without taking into consideration their relationships and the processes of their own transformations. In itself, this method is not "wrong". On the contrary, it is a necessary phase of any inquiry. In order to clarify relationships and processes, one must first get a clear picture of what actually gets involved in relationships and what goes through processes of transformation. In other words, one must first answer the "what is it" question in order to solve the "how does it act and change" problem. What leads us astray with this method is the often encountered mistake of stopping at the first result and pretending that this result is the ultimate truth provided by "scientific" research. Whereas, as shown by Hegel and brilliantly realised by Marx in his own studies, the real scientific method requires the consideration of the object under study in the cluster of its relationships and within the process of its incessant transformation. Nevertheless, metaphysical analysis is an indispensable stage of inquiry. In the context of the present research, this means that in order to fully elaborate the dialectical identity of the individual and social aspects of humankind, one must first develop the particular understanding of each aspect separately. Up until now, the most satisfactory explanation for the social aspect of this twin concept has been supplied by the historical materialist approach. Marx, Engels, and their followers have not only offered us a valuable tool-box for the analysis of social life and its development, but they have also provided the most thorough analysis of the capitalist mode of production. Yet, it is a fact that the historical materialist approach still has important gaps to fill concerning the symbolic aspect of social life, the workings of culture, and the nature and functioning of language. Moreover, coming to the other aspect of the twin concept, the individual aspect, what is undeniable is that the contributions of historical materialism have remained quite slim. It follows that to be able to carry on the "metaphysical" research required for the individual aspect of the dialectic identity analysis, in addition to probing the Marxist classics, we must reach out to the vast literature on this subject. That is why for the present study, I endeavoured to glean as much information as I could from the fields of genetics, biology, psychology, psychiatry, and especially psycho-analysis. Once the findings of the particular study on the individual aspect of humanity have complemented analysis of the social aspect, the major task remaining is to blend all this knowledge with what we have learned from evolution and the related sciences, in order to elaborate the problematic of the third level of dialectical identity analysis. As Marx's analysis of production and consumption at this third level has demonstrated, the final merging of production and consumption imposes a passage through distribution relations, i.e. class struggles and exchange relations (market relations), thus revealing the characteristic mechanisms of the capitalist mode of production. I hoped that such research would enable us to unravel the maze of relations resulting in the simultaneous unfolding of the individual and social aspects of the human species in general, as well as in the concreteness of particular societies.

To properly posit the third tier of the dialectical identity analysis, the following questions must be handled: what are the mechanisms and mediations used by social relations in order to bring about individuals compatible with these relations? What are the mechanisms and mediations used by these individuals in order to create movement, development, and eventually metamorphosis within social relations? Since in the eyes of Marxism the individual is a social being right from the beginning, the acquisition of "sociality" must at least begin with birth. This means that at least one individual must stand in a continuous relationship with the new-born. And this relation must take the form of face-to-face intercourse. The newborn can only relate through its five senses, and this imposes physical proximity. In no society whatsoever can the "socialisation" of a newly born baby be achieved without the existence of face-to-face relations. This means that the first and most important question to investigate is the nature of these face-to-face structures that ensure a steady passage from animality into humanity. Nowadays, these structures are mostly composed of families, but it need not to be so. As revealed by diverse ethnographic investigations, these "structures" can be very loose. But they must exist. And what is more, they must be able to shape each new individual as a "social" being. This means that to become human, each individual must grow up as a distinct conscious person. What makes this possible is the existence of a common language, i.e. of a shared system of references allowing intercourse. No humanity is possible without a common pool of meaning, without an all-encompassing cultural environment. Senses are not peculiar to humans. What differentiates them from animal senses is that in humans they function as intermediaries for "meaningfulness". To be human is to be able to move from sensations to perceptions, and from perceptions to feelings and thought. In other words, to transcend the biological and reach the cultural. It follows that face-to-face relations achieving the socialisation of the new-born cannot be reduced to relations of physical proximity; they must also constitute cultural relations. For physical proximity and intercourse, the five senses which we share with the animal kingdom suffice. But evidently this cannot be enough for cultural relations. To be human, to be able to function as members of the human species, we must possess "cultural sensitiveness" along with "somatic sensitiveness". How can it be possible to be part of a network of meaning-exchange without possessing "feelers" enabling us to grasp the meaning of the feelings and thoughts which make up the "stuff" of cultural, i.e. human, intercourse? Our senses allow us to perceive physical inputs. But how can we explain our sensibility to meaning-laden phenomena?

The first half of my research was devoted to finding a suitable approach for a simultaneous analysis of the twin individual/social aspects of human reality. The second was mainly tormented by the increasing realisation that human individuals could not function as a "subject" as long as their potential for communication was not clearly determined. It is true that human beings communicate via sounds emitted and heard owing to their somatic capabilities. But what about the exchange of "meaning"? Such an exchange is impossible unless there is a shared understanding. And the matter of such an understanding cannot be infinitely varied. Exactly as our bodily senses have capabilities which are qualitatively and quantitatively delimited, so must the capabilities of our souls also be qualitatively and quantitatively demarcated. If it were not so, every new human society could vary ad infinitum from any other, rendering impossible the definition of a unique human species. "Differences" can only occur on a basis of a "commonality" (e.g. you cannot "compare" a volcano and a bird). How could we compare two individuals or two cultures if their nature was infinitely different? How could we communicate if we did not realise that one feeling is about trust, one about sympathy, another about lust, etc., even when its concrete shape and content, as well as the identity of the protagonists, may vastly differ from individual to individual or from culture to culture? It is this way of thinking that led me to a meticulous inquiry in order to discover what I termed "modes of feeling". These are not concrete feelings, they are "modes", i.e. just possibilities, but possibilities defined according to their nature. There must be a limited number of them, and they must be present in every individual and every society. Just like the human senses that characterise the human body, these "modes" are the axes around which every human soul or culture is organised. The nature of these lines of demarcation must be well understood: as a bodily capability, sight is unique, but it allows us to see an infinity of various objects. We could use fear as one example of these modes of feeling: its sources, forms, impact on social relations, etc., can and do vary immensely, but there is no homo species or human society immune to fear, just as there is no normal body deprived of sight. This discovery was the breakthrough which allowed me to build up the essentials of a Marxist theory of the social individual. (Section two of the second chapter is particularly devoted to the exposition of the concept of "modes of feeling", but we will extensively encounter it both in other sections of the second and in various sections of the third chapters, as it is indispensable for the grasping of both psyche and culture.)

Section 2. Roles-Structures-Institutions. Customs/Habits

If all human actions were random and even the slightest prevision was impossible, no social science whatsoever could exist. This is true both for the individual and the social aspects of human reality. We thus need specific concepts allowing us to deal with the more or less crystallised forms of human action. As this necessity is universally recognised, social sciences broadly share a series of such concepts. These are "roles", "structures", and "institutions" for the social, and "habits" for the individual. In every society, individuals are expected to abide by certain rules when they assume particular social positions. This cluster of specific rules constitutes what is commonly referred to as a "role". Roles are usually elements of a system of relationships. These systems are more or less firmly established patterns of interaction and they regulate the reciprocal behaviour of roles. These established patterns constitute "structures". When these structures are relatively solid and longstanding they are called "institutions". All these repetitive social actions are performed by human subjects. What makes this performance possible is the fact that individuals are already endowed with "habits", which have prepared them to assume roles in order to function as parts of structures and institutions of their own society. (These already enshrined habits constitute what Bourdieu calls "habitus".)

Roles, structures, and institutions need energy to be able to perform. This explains why they cannot move unless they are activated by human individuals. Their exclusive source of social energy is to be found in individual human bodies. True, there are other natural sources of energy. But all these sources can only be enlisted by human bodies using their physical energy under the guidance of their brains. Social energy is always individual energy brought together in the amounts needed by social activity. Individual subjects, "agents" are indispensable for the functioning of social roles, structures, and institutions. Energy is always supplied by the individual. In contrast, all the material and spiritual tools, the means of action, and its forms are provided by the social. The social "moulds" individual action and furnishes it with all sorts of material and spiritual means depending on the level of development of the society in question. In sum, there are no social structures without subjects armed with a variety of material and spiritual means, and all individual subjects are themselves structured. (We will see below that the soul – the psyche – itself is a structure.)

There is no room for a "metaphysical" (in the Hegelian sense) approach in the subject–structure relationship. All the metaphysical dichotomies are superseded here. For example, in the frame of a social structure animated by reciprocal relations between individuals, the subject itself is a structure composed by one or many individuals. On the other hand, the object of the acting structure is also one or many structures. But as a reciprocal interaction is in play here, the one or many structures seen as subject/s from a certain point of view may well appear as object/s from another perspective. Again, the structure/action dichotomy is an insuperable problem for a structural approach that views structures as empty frames to be ultimately filled by individuals. Marxists do not have such a problem. As the very existence of structures depends on their activation by individual actions, their existence supposes their being activated. It follows that the problem of the activation of structures is inexistent for Marxists. This is also the case for the cause/result dichotomy. In the Marxist view, the "cause" owes its status to the fact that it begets a certain result and, reciprocally, what makes a "result" is the fact that it is brought by a cause. The pairings of compulsory/contingent and normative/fortuitous are also meaningless for Marxism if their respective terms are taken independently. From a Marxist viewpoint, the contingent is the means of realising the compulsory, just as the fortuitous is how the normative fulfils itself. What appears as "chance" to individual elements of structures is in fact a series of contingencies accomplishing structural necessities. The same goes for the normative/fortuitous dichotomy. Results which are the outcome of the laws of motion of structures generally realise themselves through an infinite number of encounters, annihilating each other or getting added to one another. Finally, the dichotomy between interior/exterior may also be easily misleading when taken too far. Indeed the relation between two structures may well be internal and external at the same time. Considered independently, the relation between these structures will be external. But when these same structures are considered as elements of a bigger structure, their relation becomes internal. Just as a factor labelled "element" acquires an internal character, whereas one labelled "condition" appears as external; and the preferred labelling will depend on our frame of analysis. But let us stop here and make our intention absolutely clear. We by no means pretend that all these distinctions are meaningless and that they should be dropped. This would be nonsensical, as Marxism itself makes use of them. This is simply a warning against a "metaphysical" use of these dichotomies, which are easily conducive to error when they are handled as if their terms were absolutely independent from one another.

Understood in this way, it is obvious that Marxism's conception of roles-structures-institutions and habits is neither "structuralist" nor "individualist". In the structuralist conception, roles, structures, and institutions exist independently from individuals. They are always "there". They can exist on their own, without necessarily having a subject. Individuals come and "fill" preformed frames. This is inconceivable from a Marxist perspective, as dead frames cannot exist on their own. They only come into existence with the concrete behaviour of individuals acting in accordance with their pre-established habits. The individualists, who hold the opposite view, clearly see that roles, structures, and institutions are made by individual performance, but they imagine that the individuals are free to create the social structure of their choice. This is also inconceivable for Marxists, as such an undetermined "abstract" soul cannot exist. A social individual is always an individual already determined. As Marx and Engels state in *The German Ideology*, "The fact is, therefore, that definite individuals who are productively active in a definite way enter into these definite social and political relations".[2] As can be clearly seen, in the Marxist view all aspects of individual action are already determined: *Definite* individuals, enter *definite* relations, in *definite* ways.

The appropriation of roles by individuals, or the insertion of subjects into various structures, is neither a random nor a fixed mechanism.

Individuals have a greater or lesser latitude in their choices, depending on the level to which the social formation has developed and on the significance of the particular role in question. For example, in definite social environments, individuals can be free to refuse marriage or parenthood. Thus, the existence of specific roles does not imply that all individuals in given situations will be obliged to assume these roles without any possibility of refusing them. Moreover, individuals are also more or less free to play a given role according to their liking; this too depends on the nature of the role, its weight in social relations, the historical conditions besetting the event, and the conjuncture. As lucidly stated by Marx in *The Eighteenth Brumaire of Louis Bonaparte,* "Men make their own history, but they do not make it just as they please; they do not make it under circumstances chosen by themselves, but under circumstances directly encountered, given and transmitted from the past".[3] In other words, roles, structures, and institutions are social expectations whose binding force depends largely on circumstances. There is never an inescapable historical obligation. Even under the most adverse conditions, people can always challenge even the most binding institutional compulsion. As this is a most commonly misunderstood trait of Marxism, it is worth repeating whenever the occasion presents itself: there is no unlimited individual freedom in the Marxist conception, just as there is no historical inevitability. The human individual disposes of a degree of freedom, varying tremendously from untrammelled freedom of choice to a nearly inescapable obligation, depending on the prevailing concrete circumstances of a given situation.

The human subject, the "agent", assuming a role is always supplied with a whole range of material and spiritual means that can be used in the performance of this role. The most important of these means is language. Language is not only a means of communication, it is also the "general store" of all knowledge about the role in question. Apart from language, every agent in every situation is always provided with a whole range of material and spiritual means, which can be used as tools to perform this specific role. In fact, the whole culture of a society, at a given moment and within the context of a specific set of relations, acts as a potential tool-box for the performance of diverse roles. Clearly, the nature, quantity, and quality of all these material and spiritual means depend on the degree to which the relevant social formation has developed and on the concrete circumstances of any social action. To sum up, in the Marxist conception of social structures, not only active subjects but also various material and spiritual means are "constitutive elements". Overlooking the agent can mean falling into the mistakes of structuralism, whereas ignoring the means may lead to the mistakes of individualism.

Section 3. Active Determination/ Passive Determination

Every social science assumes some degree of determination at all times. As underlined in the previous section, total randomness, or total indetermination, is incompatible with any social science approach. But determinations can be extremely varied: creation, destruction, delimitation, shaping, spoiling, disfiguring, sullying, exciting, inciting, provoking, conditioning, etc. During my previous studies, especially in my *Essay on the Nature of the State*, I found it helpful to give precedence to two kinds of determinations: the active ones and the passive ones. It is my firm belief that such a simplification allows us to single out what is most important in the maze of determinations that usually confronts us. An *active determination* is what constitutes the "cause" of a happening. It is the factor that accounts for the existence of a certain result. Whereas a *passive determination* is the name given to all other conditioning factors that facilitate some outcomes while impeding others, shaping the results of active determinations in multifarious ways and endowing them with their peculiar characteristics. This distinction is one of the most common in everyday life. Some examples will help to clarify it. Let us start with your present activity: you are reading this book. What made you read it? It might be an academic assignment, the advice of a friend, a critique in a journal, etc. Whatever it is, this constitutes the active determinant of your present activity. Whatever influences the result of this activity constitutes the passive determinants: the conditions in which you are reading, your state of mind, your mood, your education up to now, etc. Let us take another example: imagine you call on a friend to take them on an outing. If they accept, the active determinant of your outing is your invitation. All the rest falls in the domain of passive determinations: the means of travel (walking, cycling, driving, etc.; if you are on an island and there are no boats, this is an option denied from the outset), the route chosen, the weather, traffic conditions, etc. Or let us say that you play basketball. Why? Because you always wanted to, or to please your parents, or to accompany a friend, etc. Among all these reasons explaining your present activity, the one which has caused you to play is the active determinant, while your height, the state of the sport facilities, your time-table, your financial possibilities, etc., make up for the passive determinations.

One of the main reasons why this distinction is so useful is that the active and passive determinations may easily shift their positions. Let us return to our examples. Imagine that the conditions surrounding your reading deteriorate suddenly; a noise which was innocuous intensifies rendering it impossible to continue. Conceptually expressed, this means that a passive determination has changed into an active one, thus directly deciding the outcome of your reading. In the second example, if your ride is on bicycles, a deflated tyre will easily terminate it; in other words, it will actively determine it. Finally, if you move to a remote village, where playing basketball is practically almost impossible, all your yearnings may well be easily rendered "passive" and thus inoperative in the absence of an active determinant. It is mostly this mutability that makes the distinction of active/passive determinations so useful in the scientific analysis of concrete situations. We know that, as a rule, humans are determined by their social life. In other words, the active determinant of individual behaviour is the social relations entered by the individual in question. But let us imagine that this individual gets a nasty cancer or is involved in a serious accident leaving them crippled. In that case, it may well be that her succeeding behaviour will be mainly determined by her bodily conditions. Such a pliancy makes this particular conception of determinations incredibly practical. Let us say that, as a general rule, the active determinant of social action is the material conditions surrounding social agents. But in exceptional conditions, it may well happen that moral factors are the main driving forces behind a specific action. If we do not employ a cognitive instrument, such as the one presented here, we will end up either blinding ourselves to reality or concluding that the main tenets of our theory are erroneous. As can be seen, what is at stake here is in fact a very old, extremely basic philosophical problem: the question of the validity of exceptions. A dialectic worldview permits us to see clearly that rules and exceptions are not only both valid but are in fact indispensable to each other. What makes the active/passive distinction of determinations so valuable is that it allows us, among other things, to satisfactorily handle this age-old problem.

This cognitive tool is particularly helpful in the Marxist analysis of social facts. Let us take as an example the analysis of roles-structures-institutions in relation to individual subjects. These subjects are the active determinants. All the material and spiritual means that constitute specific roles-structures-institutions are passive determinants. These are part of the general cultural wealth of the social formation in question. As a rule, the individual subjects, or "agents", are the active determinants of social actions. But there are two important factors which should be taken into consideration. First, the agents themselves are already determined. These determinations may take the form of deeply embedded habits influenced by prior passive determinations. Secondly, material or spiritual components of the cultural environment may acquire a compelling force that gives them the strength of active determinations. As can be seen, it is not the nature of a determination which makes it active or passive. It is its *function* and *strength*. This explains why such a cognitive instrument is particularly useful in concrete analyses. For it is only within concrete analyses that we are able to properly reckon the functions and strength of social actions.

As active determinations are due to agents, and the passive ones are the result of conditions, one may easily conclude that active determinations always arise from persons, whereas the passive ones are always inanimate. That would be an error. Function and strength are independent from whether or not determinations originate in human beings. An earthquake can act as a very active determinant in one situation, while a whole crowd may be reduced to being a passive determinant in another.

One more crucially important conclusion to be derived from the distinction between active and passive determinations relates to the nature of the outcome of determinations. As already stated, the *existence* of a result depends directly on the existence of a corresponding active determinant. But the *nature* of the result depends on the passive determination. This is a very simple and basic trait of Marxist theory and the degree to which it is ignored is really baffling. Marx states that the fundamental determinant of social life is the mode of production of the basic necessities for the survival of the society in question. Taking this statement about the basic importance in the production of the means of subsistence, "thinkers" all over the world look to find "economic" results in order to vindicate or rebut Marx's thesis. However, such a search is critically misled and cannot fail to end in blunders. Indeed, the most fundamental determination of all societies is their economic life; consequently, there cannot be any aspect of social formations that is not fundamentally influenced by this basic determinant. But the result of this determination will carry an economic nature only in the sphere of the economy. In all other spheres, the result of this economic determination will *never* be "economic". If it were so, all social life would have been reduced to the economy and the words of Marx would be an absurdity. I have already stated this basic principle in my *Essay on the Nature of the State*: "The result of a determination depends on the nature of the determined, and not on that of the determinant."[4] I should have added that the *existence* of the result was dependent on the existence of the active determinant. But this being obvious and not at all disputed, I found it superfluous. I also accompanied this with the following citation from Marx, which I repeat here as it has not been sufficiently noticed: "Trade always has, to a greater or lesser degree, a solvent effect on the pre-existing organization of production, which in all their various forms are principally oriented to use-value. But how far it leads to the dissolution of the old mode of production depends first and foremost on the solidity and inner articulation of this mode of production itself. And what comes out of this process of dissolution, i.e. what new mode of production arises in place of the old, does not depend on trade, but rather on the character of the old mode of production itself."[5] Marx could not have been clearer. Here, trade is the active determinant. It brings about the dissolution of the old mode of production confronting it. Thus, this happening is the direct result of the active determinant. But the nature of the result will not depend on trade. If it were so, trade would have given the same result all over the world. This is not the case. The nature of the result depends on the *determined*, it is shaped by the passive determinations. The extent of the dissolution, and especially its outcome, "…does not depend on trade, but rather on the character of the old mode of production". We will see below that this principle can be superseded when certain quantitative limits are exceeded. But the basic principle is generally true and its importance for the understanding of the Marxist determinations cannot be overstated. When it is clearly seen that the results of determinations are shaped by the passive determinants, and principally from the nature of the determined, there is no longer any need for all these laborious developments concerning distinctions between fundamental and so-called "principal" determinations, the search for "over-determinations" and the apparently inexhaustible "explanations" of the determination "in the last instance". To give just a couple of simple examples, let us imagine that social conditions have made a certain individual go mad. The existence of madness is obviously due to these specific social conditions. But madness itself is a somatic phenomenon, it takes a *bodily* form and it will have to be tackled as such. Or let us imagine that a riot has provoked the burning of a building. The cause of the fire is obviously social. But its “nature” is *physical*, and it will be put down with the help of physical means.

It is as simple as that, and it is vitally important, especially in concrete analyses. The concrete takes its specificity from the diversity of its determinations. Knowing that everything depends fundamentally on the mode of production does not take us very far. We have to analyse determinations in all sorts of fields, and often in detail, to be able to decipher reality and to modify it.[6] As already stated when considering the usefulness of dialectics, the only way to use it efficiently is to become knowledgeable about the particular factors relevant not only to the concrete active determination, but also to the passive ones.

Section 4. Conversion of Quantity into Quality and Vice Versa

The metaphysical approach fails to acknowledge that quantitative changes within objects provoke qualitative changes that substantially modify the object in question, and this is one of the main reasons for the damage it does to scientific thinking. The history of sciences is filled with various examples of this methodological blindness. For example, the disregard for the fact that beyond a certain level of development, the severance of humanity from animality is totally inevitable. As will be dealt with later (Chapter II: section “The Evolutionary Origin of Feeltypes”), carried by his impatience to prove that humankind is a product of evolution, Darwin repeatedly exposes the continuities between humankind and the most advanced representatives of the animal kingdom. He insistently singles out diverse instances of tool usage, modes of communication, socialisation processes, and manifestations of feelings among animals. However, he does not realise this simple fact: beyond a certain threshold, there cannot be any compatibility between properties that are specifically human and being part of the animal kingdom. An animal species exhibiting these particularities can no longer be considered as a kind of animal. Its conversion into humanity must be recognised. It is true that humankind originates from animality, in evolution. But it does not follow from this fact that humankind itself is a kind of animal, even if of an extraordinarily developed kind. Beyond a level of quantitative accumulation, the object under scrutiny changes its nature. To maintain that the objects remain unchanged is conducive to methodological blindness.

Again, a cultural transmission of knowledge relative to certain means of survival is encountered among certain animal species. A lot of animals teach specific behaviour to their offspring. In other words, a certain amount and variety of knowledge and capability is culturally transmitted. But this mode of cultural teaching is nowhere predominant in any animal species. When it gains prevalence, we are no longer faced with an animal species, we come face to face with humankind.

Animals become adapted to the environment through modifications in their anatomy and physiology. This is the law of evolution. There is no doubt that this law keeps exercising its effect after the appearance of humankind. But with the emergence of the human species, the environment decisive for survival becomes the social environment. Beyond a certain accumulation of social traits, natural environment metamorphoses into social environment. The foremost means of evolution, organic adaptation, becomes the consequence of social life.

There is a plethora of other examples for the conversion of quantity into quality. Let us consider individual actions. Below a certain concentration, these may appear as a matter of private fancy or even as a dream of a distant utopia. But once a certain threshold is reached, these same individual actions are transformed into revolutionary activity. Quantity alone may change the nature of individual initiatives, transforming them into components of a revolutionary uprising.

The same goes for the passage from passive determination to active determination, as considered in the previous section. Once a certain intensity is reached, a passive determination gains the power of deciding the outcome, in other words it gets "activated". The role of the body in human behaviour is a good example. As a rule, the body is a passive determinant. But when it is seriously impacted upon, when for example it is confronted with a serious ailment, this normally passive body may well impose a definite outcome. Let us take the example of a prime minister who resigns from office. In normal circumstances, this would be the result of a political setback – an electoral defeat for example. But it may well happen that the person in question may have to resign due to serious illness. Then we have a passive determinant getting active. This passive determinant can even be quite trivial in normal conditions, as is the case for the venue where a certain social activity takes place. But imagine that a fire breaks out. Immediately this insignificant determinant becomes the most actively compelling one.

Another example concerns the rule indicated above, according to which the nature of a result depends on the nature of the determined. This rule, which holds true in general, becomes inoperative when a certain threshold is crossed. Let us say that you slap a series of individuals in the face. Each individual will react differently according to their own character: one will hit back, another will shout, another will flee, etc. Clearly, the outcome of the same action will differ according to the character of the person attacked. But imagine now that you run over these same persons while driving a truck whose brakes suddenly begin to fail. There will scarcely be any room for distinguishing within the heap of squashed bodies. As seen, beyond a certain point, passive determinations lose their efficacy and only the active determinant decides what the outcome will be.

Finally, it must be noted that the conversion between quantity and quality is not one-sided. In other words, quality may have a decisive impact over quantity. Take the example of a technical novelty. This technological breakthrough may provoke an unequalled increase in production. This is what we normally witness in the course of economic development.

In summary, quantity and quality are not indifferent to each other. It is only a metaphysical obstinacy which can blind us to this simple fact. It follows that a dialectical enquiry must always keep an eye on the quantitative, in order to be able to spot qualitative changes in the object under scrutiny.

Section 5. Level of Abstractness/Concreteness

Marx has written very few texts directly on method. This could be seen as a good thing. I particularly have in mind the Preface to *A Contribution to the Critique of Political Economy* (1859), which has been the source of incredible misinterpretations of Marxism. And unfortunately many of these misinterpretations are still widely held. As it will be well remembered, in this text Marx divides society into an "economic structure" and a "superstructure". In his own words: "In the social production of their life, men enter into definite relations that are indispensable and independent of their will, relations of production which correspond to a definite stage of development of their material productive forces. The sum total of these relations of production constitutes the economic structure of society, the real foundation on which rises a legal and political superstructure and to which correspond definite forms of social consciousness. … At a certain stage of their development, the material productive forces of society come in conflict with the existing relations of production, or – what is but a legal expression for the same thing – with the property relations within which they have been at work hitherto. From forms of development of the productive forces these relations turn into their fetters. Then begins an epoch of social revolution. With the change of the economic foundation the entire immense superstructure is more or less rapidly transformed."[7] Two pages all in all, to summarise the richest conception of social life ever elaborated up to this day! And it is a real tragedy that a kind of "digest" has been construed out of these cursory statements. It reads as follows: human activities are carried out independently from one another. The most important of these activities are those devoted to the development of productive forces. These have a pace of their own and they determine a specific set of relations of production, which correspond to the level of development achieved by the forces of production. These relations of production too are independent of human will. At a certain stage, an automatic clash occurs between the forces of production that develop naturally and the relations of production which envelop them. Thus, an epoch of social revolution begins. As a result, more or less rapidly, the whole superstructure is transformed so as to comply with the new relations of production corresponding to the developed level of productive forces.

Unfortunately, this impoverished credo, identifying Marxism with a most mechanist technological determination, predominated the Second International after Engels' death, and the third International after Lenin's death. This vulgar interpretation of Marxism is still the most currently held, both in academic and lay circles.

But fortunately, Marx left us another text on method, which contrary to the preceding one, was not published during his lifetime. It was written in 1857, as part of the manuscripts later labelled as the *Grundrisse*. In the introduction to this work, Marx discusses "The Method of Political Economy". As he warns us there, at first sight the right approach to political economy would seem to be a study of the population, as it is the most concrete basis of economic life. But this would be misleading as the population is a mere abstraction if the classes composing it are not taken into consideration. But classes themselves are empty notions if we ignore the elements on which they rest, wage labour, capital, etc. And these in turn presuppose, exchange, division of labour, value, money, prices, etc. Thus, in the course of our investigations, we are necessarily driven to the study of more and more abstract notions until we reach the simplest determinations. This is how analysis of social reality should be conducted. When these simplest concepts have been rounded up, proper theoretical elaboration can proceed. The task then is to reconstruct the concrete, step by step, following the development of the appropriate determinations up to the point of reconnecting with the population. But now this concept is no longer an arbitrary construct, a simple empirical phenomenon. It is a truly concrete concept: a scientific mirror of reality.

As can be clearly seen here, we are far from making sweeping generalisations about social reality. On the contrary, we are led into a delicate, intricate method of social analysis, which can uniquely validate a truly scientific mode of exposition. To summarise it in one phrase: the right path consists of digging in the apparent concreteness in order to lay open, step by step, the determinations hidden underneath, up to the simplest determinations; and from this point on, to elaborate the theory of the social object under consideration, by bringing into play more and more concrete determinations up to the theoretically most concrete reconstruction of social reality.

When Lenin in his effort to determine the most valid revolutionary strategy calls for "the concrete analysis of the concrete situation", it is exactly this theoretical approach that he has in mind. Firstly, the simplest and most abiding determinations must be discovered. This is how he reaches the conclusion that the determining fact of Russian society at that stage is its domination by the capitalist mode of production. He is able to devise a realistic revolutionary strategy by starting from this fundamental determination and progressively adding the specific determinations of Russian society: its location in the periphery of imperialism; its population vastly dominated by the peasantry; its multinational composition under the sway of Russian chauvinism. He further refines this strategy by taking into consideration the wholly new situation created by the world war, thus forging the most theoretically concrete tool for enabling him to lead a successful revolution. Lenin is the most practical revolutionary agent, because he is the most theoretically correct. It is the right interplay between abstractness and concreteness which is the key to his success.

The method of political economy exposed by Marx offers us two main principles concerning the proper handling of abstractness/concreteness. The first can be labelled as "systemic consistency" or "principle of totality" in a Hegelian sense. The second concerns itself with the congruence of the levels of abstractness/concreteness dealt with in any analysis or theoretical elaboration. But these labels are most unimportant. What counts is their content.

The first principle requires us to be most attentive to the systemic boundaries in our theoretical and practical investigations. Objects under study must be handled as elements of the system to which they belong. To take an example given by Marx in the third volume of *Capital*, the merchant's capital, which is much older than capitalism proper, presents a different nature according to the system under which it operates; merchant's capital is different in the ancient world, in the feudal mode of production, and in capitalism.[8] An abstract or concrete analysis ignoring this historical fact is bound to lead to error. Any element of a system can be properly evaluated only within the structure of its system. Analysis of the empirical concreteness of elements of this system, with a view to reaching the simplest abstract determinations underlying them, can only be properly conducted if, and only if, the logic of the system is constantly kept in mind. This holds as well for the theoretical elaboration starting with the simplest determinations and aiming at a theoretically concrete reconstruction of reality. If one overlooks this principle and studies any element independently, such an abstractness or concreteness can only have an illusory object.

Marx emphatically warns us against considering elements of systems as having an independent history on their own: "It would therefore be unfeasible and wrong to let the economic categories follow one another in the same sequence as that in which they were historically decisive. Their sequence is determined, rather, by their relation to one another in modern bourgeois society, which is precisely the opposite of that which seems to be their natural order or which corresponds to historical development. The point is not the historic position of the economic relations in the succession of different forms of society. … Rather, their order within modern bourgeois society."[9] This is also a warning against analogy. For analogy can easily overlook systemic differences and concentrate solely on the object taken independently from its relationships.

The second most important principle has already been touched upon in the introduction and in the first section of this chapter, as I consider it a most crucial methodological principle. It has to do with strictly respecting the borders of abstractness/concreteness for the level at which we operate when conducting any study. Both in the step-by-step analysis leading from the empirical concrete to the most abstract determination, and in the theoretical elaboration starting from the most abstract, consistency in the level of abstractness/concreteness must be most meticulously observed in order to reconstruct the theoretically concrete. Inobservance of this basic principle is bound to lead to sweeping generalisations or to baffling short cuts that are equally woeful.

To properly understand this principle, one should keep in mind the identicality of abstractness and generality on the one hand, and of concreteness and particularity on the other. No scientific thinking allows jumping from the general to the particular and vice versa. The necessity of intermediary stages, of inescapable mediations, must never be overlooked. Let us take the example of the fundamental determination in the Marxist conception: determination by the mode of production. Such a determination will almost never be direct. It will always exercise its effect through concrete forms of class struggle, raging through all kinds of social, political, and ideological intercourse. Therefore, sorting out its effect will require a thorough investigation of the various transformations undergone by the same determination throughout its operation.

To sum up, in dialectics generality can only be built through proper specificity, as concreteness can only be the result of sound abstraction.

Section 6. The "Sociality" of the Human Individual

According to the Marxist conception, human individuals themselves are social structures. Like any structure, the individual too is composed of active and passive elements. The active element of the individual is the soul (the psyche), which is itself a social product. Just as the active element (the "player") within social structures is the individual performing their role, the active element (the "player") within the individual is their psyche. Again, as is the case with the individual performing their role, this psyche is always concretised by a body. When the individual is considered as a structure, the elements materialising its active element, i.e. the psyche, and serving it as means can be listed as follows: a living organism, a body; a genetic structure, a genome which determines the biological properties of this body; features added to the body by the care given to it during gestation and through education starting with birth; material and intellectual potentialities, capabilities both innate and nurtured thereafter; a certain language; a basic temperament (e.g. exuberant, daring, jovial, calm, shy, sluggish, etc.) being the combined product of genetic inclination, of conditions surrounding gestation, and of the impact of face-to-face relations; modes of behaviour shaped by playing the role of a "child" as imposed by traditions and ideologies, internalised in a specific way by a determinate family; modes of conduct adopted by the young individual first as unconscious reactions and mimicry, and later as conscious preferences (these may also be learnt from people outside the family, from television, films, plays, books, etc.); a specific character formed by all these elements, these active and passive determinations, which are combined through this process of formation, leading to an original synthesis (a personality) compelled to retain its structural integrity in any situation.

What must be particularly stressed according to the Marxist conception of the human individual is the fact that *all components* listed above as elements of the individual structure are *social products.* This is not only true for variables destined to instil definite habits in the young human after birth, variables for which the social character is obvious, but it also holds for variables appearing to be absolutely organic. To start with, the human body is not an organism like any other, it is an organism produced by social evolution. Therefore, in contrast to any other animal body, it is a social product. Undoubtedly, like any other organism, the human body cannot survive without a steady give-and-take material connection with its environment. But what the specificity of the human body comes from is the fact that all these material connections are regulated by social norms. Effectively, all the practical activities of the human body, from its food intake to its discharge, from its work to its sleep, from its courting to its mating, from all of its sensual pleasures to its sufferings, i.e. all the aspects of individual existence, are realised through social relations. This is true for the species as a whole and for any individual in particular. Not only is the human genome generic, but each particular genome is the product of social intercourse. And what will become actualised in this genome will in fact depend on the social relations of each individual. The social constitutes the very being of humanity. Thus, the articulation of the individual with the social is a *non-existent* problem. You cannot "socialise" what is *already social.* Marx gets to the crux of the matter when he states most simply (and the stress is his own) that "The individual *is the social being*".[10]

As this question is often very misunderstood, it is worth delving into it a bit longer. There is no doubt that like any object the human body too can be analysed down to its tiniest components, be they its chemical elements, atoms, or even its sub-atomic particles. But no one will infer from this fact that individual behaviour can be explained out of the atomic structure of any individual's body. It is no doubt true that a malformation in the atomic structure of cells can provoke very serious illnesses resulting in a major impact on individual life. But it certainly does not follow from this fact that one can explain a personality by analysing the atomic structure of the body. This specific organic being is essentially a means of "sociality". For example, the matter out of which a table is made is some kind of raw material. Wood can be easily replaced by glass, or concrete, etc. As long as the object functions as "a table", the stuff used for its construction does not count. Of course, there is not (yet?) any alternative to the organic entity currently functioning as a human body. Nevertheless, this fact in no way allows us to consider the organic as the key to the elucidation of the human.

The active element of the individual structure is the soul (the psyche). And the soul is inherently social. It is a social object. All the other elements composing the individual structure are basically means serving the soul. It is certainly true that in exceptional cases one of these passive means can become an active determinant. Just as cancer developing in someone's body becomes the decisive factor in their life, any one of these means can acquire such a determining strength. As indicated before, there is nothing mysterious in the transformation of quantity into quality, in the "activation" of a passive determinant as a result of its exceptional strength. This applies to individuals born with exceptional capabilities, for example a musical genius. When a person of this calibre encounters the possibility to train their genius nothing will be able to stop them. This will be the main determinant of their life, and what is the active determinant for the vast majority of people will remain ineffective in their case. But obviously this is an exception. The existence of such rare exceptions, which can affect any of the means constituting individual structures, cannot invalidate the basic truth regarding the fundamentally social nature of individual structures.

Notes

1. "The Method of Political Economy". Marx (1973, pp. 100–108).
2. Marx and Engels (1998, p. 41).
3. Marx (1852, p. 96).
4. p. 21. As already indicated this text is accessible through this web address: cemerogul.wordpress.com
5. Marx (1991, p. 449).
6. Incidentally this provides the logic of the necessity of elaborating a specific theory for each different dimension of social life: a theory of families, a theory of gender, a theory of language, a theory of literature, a theory for each art, etc. A theory limiting itself to the exposition of the active determination by the mode of production would indeed remain quite short if it

were not every time supplemented by the specific theory explicating the functioning of the passive determinations particularly relevant to the object under study.

7. Marx (1859, pp. 181–182).
8. Marx (1991, pp. 442–443).
9. Marx (1973, pp. 107–108).
10. Marx (1988, p. 105).

References

Marx, K. (1852). The Eighteenth Brumaire of Louis Bonaparte. In *Karl Marx and Frederick Engels Selected Works in One Volume* (pp. 94–179). Lawrence and Wishart, 1970.

Marx, K. (1859). *Preface to a Contribution to the Critique of Political Economy*. In *Karl Marx and Frederick Engels Selected Works in One Volume* (pp. 180–184). Lawrence and Wishart, 1970.

Marx, K. (1973). *Grundrisse*. Random House, Inc.

Marx, K. (1988). *Economic and Philosophic Manuscripts of 1844 and the Communist Manifesto*. Prometheus Books.

Marx, K. (1991). *Capital Volume Three*. Penguin Books in association with New Left Review.

Marx, K., & Engels, F. (1998). *The German Ideology*. Prometheus Books.

Chapter II: The Soul (the Psyche)

Section A: Theory

1. The Soul as the Individual

The soul is the *individual* manifestation of culture, or in other words, society. When we look at each other, it is really the soul that we see. When we meet someone we know, they have a name and a meaning: they are a person. But even when we meet someone we do not know, they still convey a *meaning*: fellow citizen, foreigner, woman, man, elderly, young, venerable, ridiculous, frightening, trustworthy, etc. A person we do not know is, nevertheless, *a person* with whom we are somehow "acquainted", because they are one of our species – a fellow creature as people used to say. They are completely different to a non-human entity, an animal, a plant, or any object. The way we see each other is always as a feeling-thought combination. Our own feeling-thought combination "understands the language" of the other feeling-thought combination we are confronted with. This is the common language of humanity, which encapsulates a multitude of specific languages. It is the element that gives meaning to our human exchanges. It is the language of the soul.

The secret of being human is this continuous exchange, and our entire human configuration depends on it: "if you lie down with dogs you get up with fleas", "one rotten apple spoils the barrel", "one rotten egg spoils the whole pudding", "know a man by his friends". The copious maxims passed

C. Eroğul, *Marxism and the Individual*,
https://doi.org/10.1007/978-3-030-83662-7_3

down through generations suggest that our ancestors had already grasped humanity's defining reality many years ago.

Every individual juvenile develops through face-to-face relationships. There is no other choice. The human empathy gleaned from this process gives each individual a sensitivity to affective patterns that emerge in a range of disparate times and places. We are constantly being shaped by our experiences, and constantly acquiring new sensibilities that allow us to empathise, not just with those alive now, but also with past humankind; contemporary Shiites still sense the pain of a murder that happened over 1300 years ago at Karbala, and today's revolutionaries still hear the screams of the Paris Commune, about 150 years after gunfire decimated the insurgency.

The contagiousness of emotion is compelling evidence of the collective nature of the soul. One of the defining characteristics of the human individual is our ability to be infected by the laughter of others, share the tears of those who are crying and the fears of those who are afraid. This is an empathy created by face-to-face relationships. Furthermore, our capacity to abstract ourselves from the concreteness of feelings gives us the ability to extend this empathy beyond face-to-face relationships. Complete strangers acting out a drama on stage or screen can invoke mutual feelings in us, just as we also share in the same emotional journey of stories we absorb in oral or written literature. Therein lies the secret of art.

This rule applies to every type of emotion. For example, we are aroused by the sensuality of others; most people are aroused by seeing others engage in sexual activity or begin to feel hungry when they see others eating. Every emotion calls out to us. We join others in leisure activities or get involved in fights with them. Moreover, sharing emotions makes them even more powerful. When drawn into situations like war, massacre, ritual, or celebration, individuals exhibit the kind of extreme behaviour that they would never contemplate alone. Humankind, in all its aspects, is *infectious*. Because, essentially *we are one*. Our essence is others. What makes a person who they are is always *an other*.

In all species, the first requirement of "knowing" another is being able to attach meaning to certain data, in other words to see this data as an individual "sign". Some signs are specific to a certain species. Bees understand each other's movements and circling patterns (dances). A noise emitted by one of the species tells others about nearby food, and a different sound relays information about the type of approaching danger. Naturally, for the news to be passed on, the sound must be comprehensible to others of the species, i.e. it must function as a sign.

Like animals, humans only get along with each other by accepting certain "happenings" or events as portents, such as a roar signifying the approach of a lion or a crack of lightning warning of imminent rain. For a happening to be a sign is tantamount to saying that it *carries a meaning*. The members of every species communicate with each other using signs unique to their species. It is only by assigning mutual meanings to these signs that the individuals of the species are able to send each other specific messages. It is the specificity of the human species that has enabled them to enhance the universal (abstract) level of these signs, transforming them into individual symbols. A symbol is a sign that has attained a certain level of universal abstraction (the transformation of quantity into quality). In this sense, it is not wrong to describe humans as a species that uses symbols. What renders thought possible is the symbolisation that enables it to be plucked from the concreteness and uniqueness of each object. Signs take on the characteristic of symbols when they become distanced or freed from the things they signify. Eventually, a level is reached at which symbols too lose their natural ties with the realities they allude to and end up in an entirely artificial relationship with the things they signify (a new phase of the transformation of quantity into quality). The symbol turns into an indicator. Language is a series of such indicators. Therefore, removing the confines of thought can make the act of thinking into a potentially boundless activity.

Not long after birth, human juveniles begin to perceive the feelings they are experiencing as signs of a certain affective state. If the sign relates to a situation that satisfies needs, relaxes, gives comfort or gratification, the juvenile tends to maintain and repeat the present interaction. If the sign is in response to a frightening, discomforting situation that provokes anxiety, the juvenile tries to escape from the present interaction and avoid it in future. The baby processes particular signs into symbols through constant repetition. For example, a smile comes to symbolise a positive situation and a frown a negative one. The conditioned reactions, consistently provoked by symbols that the child has learned to recognise, form the building blocks of the juvenile psyche. Based on the feelings it is experiencing, a child absorbs signs for the things that bring happiness or reduce anxiety, comprehending them within its own interactions and thereby acquiring the appropriate behavioural patterns. In short, interactions that give rise to pleasure, relieve anxiety, or the exact opposite are learnt by every individual in the form of feeling symbols.

Clearly, the building blocks of the psyche are built within inter-individual relationships. Symbolisation is the mechanism by which thousands of moods are abstracted from their own concrete singularities and transformed into psychical elements. As a juvenile accumulates these experiences of signs, it begins to perceive them as *affective symbols*. In other words, it begins *to make sense of* its own emotions. One type of relationship is tagged as emotionally rewarding, another as painful. The individual's behaviour is adjusted accordingly.

Making sense of one's own feelings undoubtedly does not mean consciously analysing them. This function can develop later. To make sense of feelings means perceiving what it is that they indicate. In keeping with the nature of the psyche, all signs-symbols-indicators occur in one, several, or all of its various dimensions-axes at any given time. The moods experienced obtain *affective meaning* according to the psychical *axis-dimension* on which they occur. Every feeling – happiness/sadness, trust/fear, comprehension/incomprehension, etc. – takes place in either one, several, or all the affective-intellectual axes of the psyche.

The word "feeling", as used here, must categorically not be understood as the opposite of thought. On the contrary, in humans every feeling is mixed with thought and every thought mixed with feeling.[1] Even the most emotional of feelings acquires intellectuality the minute it steps over the threshold into realisation. Likewise, even the seemingly most abstract thought inevitably has an affective impact on the thinker, who after all is neither a robot nor a machine, but rather an organism that transmits thought via a neural system. The above observations on the process by which feelings obtain meaning via signs evidences the intrinsic unity of feeling and thought. The first step to acquiring a feeling as a *specific feeling* is the process of designation (accepting it as an indicator of a certain thing), which is also the first step of recognition (of knowing and therefore thinking). Simply put, there is no such thing as unadulterated thought. Throughout this book, it is crucial to perceive of what is called *thought* as a composite phenomenon with both an affective and intellectual character.

As feelings perceived in this way accumulate on a particular axis, they are constantly building up their own axes as well as other dimensions of the same psychical system. The internal balance of each axis and its relationship to the balance of the other axes determines the dominant characteristic of each system, in other words, the *personality*. This attribute is never unadulterated and there are always other characteristics that conflict with the dominant one. The integrity of the inner balances is also very changeable. Compatibility goes hand in glove with incompatibility, incongruence, and conflict, i.e. the psychical sequence is always to some degree both changeable and contradictory. However, in order to maintain its systemic integrity, it must keep its elasticity and inner conflict down to a certain level, otherwise it would fragment, resulting in the death of the psyche, or at the very least impairment or damage.

The affective origin of each axis varies from one culture to another. In other words, the sign-symbol-indicator of every feeling is *culture-specific.* Feelings cannot be abstractly perceived and always originate from a certain relational situation. Every culture formulates its own particular templates for these relational situations. Moreover, the *mode of experience* for every feeling is also designated by the culture. With whom and how one is pleased, who will be loved and how, who will give pleasure and how, who will be trusted and how, who will like who and how, who will see the world in a certain way and with whom – all are shaped by each culture's specific templates. The same thing is undoubtedly true for all patterns of feeling and their degree of intensity, all the way from one extreme of each axis to the other.

Therefore, the concrete features of every psyche and the form that each dimension takes are determined by culture. Its *concrete origin and content* vary from one culture to another. *The universal properties of the psyche can only exist in the form of cultural specificities.*

Affective-intellectual signs-symbols-indicators are always perceived *organically.* The individual always feels them in their body. Pleasure, joy, affection, trust, the feelings of being liked, and degrees of consciousness are always body-mediated. What shapes this perception and expression is the indicator that the respective culture has internalised for this feeling. It is not exclusive to each person; it is exclusive to the culture. Without this, there could be no social relationships. If others failed to comprehend which emotion a particular behaviour, look, tone of voice, etc., indicated, people would be unable to establish relationships with each other.[2] All this is processed by a comprehensive feeling-thought language, far beyond the narrow sense of language, that acts as an intermediary for all associations and personal exchanges. The indicators are supplied by the culture and actualised by its individuals.

For the most part, people execute these exchanges via body-to-body relationships. If a feeling can only come into existence through a bodily mediation, *somatic exchange* is inextricable from affective exchange. This exchange may be from afar, as is the case with a look or a call, or it may also be indirect, such as via correspondence; but it is *essential.*

2. The Theory of Feeltypes

Humans experience a multitude of feelings in their relationships with each other, their environment and their own inner world. Moreover, these feelings change from culture to culture. This very diversity may seem to destroy any hope of meaningful classification, but on further consideration that is not the case. If there was an infinite variety of feeling-thoughts experienced by humans today and throughout history, there *could not be* such a more or less constant web of social interactions that define humans as a species.

Feeltypes are psychical *sensibility* axes. The various feelings of each culture are located on one of these axes. The feelings may be concretely very different, but their existence, people's ability to experience them, is dependent on the *existence of people's power to perceive them*. Just as the body has five different senses, regardless of concrete content, the psyche has six feeltypes: (1) Like/Dislike, (2) Trust/Distrust, (3) Approval/Disapproval (4) Joy/Sadness, (5) Pleasure/Pain, (6) Comprehension/Confusion. There are *altogether six fundamental* feeltypes, just as there are two arms, two legs, ten toes, ten fingers, and five senses.

Despite the similarity in name, we should note that these are *not specific feelings,* but virtual states (feeltypes). The way these states are named here may be unsatisfactory, as it is *almost impossible* to find names of universal validity. The reason for this is clear: these universal sensibilities are always described in culture-specific vocabulary. The important thing is to see which type of affective effect each axis corresponds to. In reality, every culture actualises these axes *in the form of concrete affectivities with different names.* The only way to analyse this multiplicity is to figure out which basic psychical sensitivity *each concrete feeling* relates to. To better understand the nature of each main axis, we can start by investigating which concrete feelings they correspond to in our own culture.

Like/Dislike within their relationship with another, someone might feel: like, love, fancy, affection, attraction, infatuation, devotion, longing, lovesickness, etc. The common thread of all these feelings is that an individual derives satisfaction from their relationship with another. At the other end of the same axis, someone might feel: hate, disgust, revulsion, annoyance, resentment, jealousy, spite, bitterness, intolerance, offence, envy, anger, fury, etc. Here the common thread is dislike of another.

Trust/Distrust within their relationships with another, someone feels: trusting, reliant, believing, dependant, have confidence in them, have faith in them, count on them, etc. The important thread is a feeling of trust derived from exchanges with another. At the other end of the same axis, someone feels: apprehensive, wary, frightened, terrified, paralysed with fear, nervous, scared to death, panic-stricken, horrified, intimidated, etc. In short, a person feels the necessity to avoid another for their own safety.

Approval/Disapproval within their relationship with another, someone feels: admired, endorsed, liked, praised, respected, valued, accepted, important, commended, appreciated. At the other end of the same axis, a person may feel: embarrassed, ashamed, mortified, crushed, disgraced, contrite, humiliated, belittled, shamefaced, etc. One of the most stressful human feelings is that of being rejected or excluded.

Happiness/Sadness within their relationship with another, a person feels: joyful, comfortable, delighted, glad, relaxed, proud, content, enlivened, cheerful, thrilled, overjoyed, full of the joys of spring, on cloud nine, ecstatic, elated, excited, satisfied, consoled, etc. The common thread of all these is "feeling good" as the result of a relationship. At the other end of the same axis are the opposite feelings: sadness, anxiety, regret, tearfulness, sorrow, anguish, distress, dejection, hopelessness, unhappiness, depression, mournfulness, misery, grief, melancholy, down in the dumps, sick to the stomach, etc. The common thread here is "feeling bad" as a result of a relationship.

Pleasure/Pain within their relationship with another, whenever a person meets or thinks about someone else, they feel: pleasure, elation, inner warmth, glowing, they want to jump for joy, etc. The relationship is the source of great contentment, happiness, and passion, felt explicitly in the body of the person concerned. At the other end of the same axis is an anguish that causes mental torment. The person feels pain, anguish, misery, intense anxiety, broken in two, bereft of joy, in bits, miserable, etc. In comparison to the other axes, emotions on both the positive and negative extremes of this axis are more intensely felt in the individual body.

Comprehension/Confusion within their relationship with another, a person feels: interest, curiosity, comprehension, discovery, perception, fathoming out, catching on, deduction, realisation, awareness, enlightenment, cognizance, learning, desire to emulate, etc. Of all the feeltypes, this is the one most associated with the mind. At the other end of the same axis is the feeling of not understanding what is going on, evoking a sense of: incomprehension, confusion, ignorance, bewilderment, obscurity, strangeness, feeling lost, getting the wrong end of the stick, unable to make head nor tail of it, etc. This sense of being in the dark leads to feeling anxious, overwhelmed, or crushed.

Except for these six axes, there are *no other* mood modes (feeltypes) that face-to-face relationships can create in an individual. These are the a priori templates of all psychical inputs (like the five senses) that an individual can experience as a social being. Doubtlessly, their content changes according to different societies, cultures, roles-structures, and individual lives. But as with the senses of the body, the template itself does not change.

Just as the range of different motions that can be performed by any body is limited by its senses, reactions, and physical skills, the range of feeling-thoughts that a psyche can experience is also limited, regardless of concrete content, form, or intensity. The capabilities of the body's senses and its available mechanisms are limited, therefore the motion types it can perform are *restricted* by its very nature. Similarly, when using any of the feeling-thought possibilities of its own feeltypes, the experiences of the psyche are limited to: Like/Dislike (enjoyment, attraction, being in love; anger, hatred, disgust, etc.); Trust/Fear (believing in, feeling safe, protected, having confidence in; fear, terror, etc.); Approval/Disapproval (proud, respected, self-assured; humiliated, ashamed, etc.); Happiness/Sadness (joyful, elated; misery, depression); Pleasure/Pain (music to the ears, mouth-watering, aroused; gagging at a smell, eardrums shattered by noise); and Comprehension/Confusion (discovery, recognition, joy of learning; confusion, stumbling in the dark, out of your depth, etc.). In summary, although feeling-thoughts come in a multitude of shapes and sizes, ultimately they are all located on *one of these six axes.*

As explained above, each one of these six basic feeltypes comprises of feelings that may be *polar opposites*, but they are on the *same axis* because they have the same basic quality. For instance, love and hate may be opposites, but they are both on the axis concerned with sensibility to affection. Therefore, it should not be assumed that all feelings on the same axis have the same orientation. *Opposing feelings* located on the same axis are frequently experienced *mixed together* to some degree. Equally, while the axes are separated from each other for the purposes of abstraction and diagnostics, the feelings they comprise of are experienced concurrently *with those on other axes,* as well as with those on the same axis. Taking the example of a love relationship again, it may be mixed with hate feelings located on the same axis; and it is also highly likely to be tangled with feelings along other axes, such as approval, pleasure, curiosity. *In real life, feelings (and the thoughts that they are inseparable from) are always mixed.* The myriad of feelings which are interspersed along each axis occur mixed with other feelings that are interspersed along each other axis. In other words, every feeling-thought experience is a *compound.*

When listing the basic feeltypes, we should not fall into the trap of seeing quantitative differences as separate characteristics or types. For example, the feeltype that includes everything from a slight recalcitrance to seething with anger is still the same type, i.e. dislike. Levels of intensity *affect the feeling but not the type.* (An analogy may help to understand this: a pleasant melody and the sound of a bomb explosion are very *different* sounds, but both are perceived with the *same* sense of hearing.) Considering them each as different types would lead to an endless series of feeltypes, resulting in the differentiations becoming blurred.

To fully understand the concept of feeltypes, they must be clearly differentiated from other psychical features such as emotion, temperament, attitude, character. *Emotion* is the experience of any feeling together with certain somatic-functional changes. It can relate to a concrete feeling that enters *any of* the six feeltypes. For example, one can feel emotional about seeing a lover, or a scientist might be emotional at the conclusion of a long-standing research project. The feeltype to be primarily associated with the first example is *Joy/Sadness* and possibly *Pleasure/Pain*; the feeltype in the second example is *Comprehension/Confusion.* (The feeltypes involved in the first emotion could possibly play a role in the second emotion as well.)

Temperament refers to a person's innate psychical characteristics, such as being cheerful or ill-tempered. These characteristics affect the *form* in which *all* the various concrete feelings belonging to the feeltypes are experienced. *Attitude* is the general mode of reaction shown by an individual in concrete life in response to stimuli from their own culture. It can concern *all* feeltype areas. For example, disinterest is an attitude. *Character* is the consolidation of attitudes into a more or less settled form. For example, indifference is a character trait; it covers *all* feeltypes.

As seen, just like the body's senses, the term "feeltype" describes an *invariant* structure. The body and the psyche encounter an endless flow of stimuli; in the body, sensory information enables us to establish the entry of these stimuli via five separate "doors", while in the psyche, the stimuli always consist of affective-intellectual elements. Feeltype information tells us that everything in this affective-intellectual tangle belongs to one of six axes. Let us say we are filled with hate or overjoyed by something we have seen and heard. The sensibilities that *enable* these concrete experiences in either of these states of heightened arousal are: the same *sight* and *sound* senses in the body; and the same Trust/Fear and Pleasure/Pain feeltypes in the psyche. These sensibilities certainly do not determine how the feelings are experienced, they simply *enable them to be experienced.*

These sensibility axes, which provide the medium for the human psyche, take *different forms* in every culture. Every culture has its own view of who will be loved and how, why someone should be ashamed, etc. A sensibility that is basically identical is experienced through *a multitude of different concrete feelings* from one culture to another. This diversity should not mislead us.

In one study of emotion concepts in Turkish,[3] states that qualify as "feeling, excitement or emotion" were listed as follows (with decreasing frequency): affection, pity, hatred, happiness, love, sadness, joy, shame, fear, longing, anger, anxiety, jealousy, rancour, flurry, hope, suspicion, respect, bewilderment, violence. We can group these concrete feelings, according to their feeltype, as follows: Like/Dislike – affection, pity, hatred, love, longing, anger, jealousy, rancour, violence; Trust/Fear – fear, anxiety, flurry, suspicion; Approval/Disapproval – shame, respect; Happiness/Sadness – happiness, sadness, joy, hope; Comprehension/Confusion – bewilderment. From among these, love, jealousy, rancour, and violence could also belong in the Pleasure/Pain feeltype, depending on the concrete situation and intensity of feeling.

As illustrated, feelings are concrete, while feeltypes are basic templates distinguished from each other by the concrete feelings they comprise. They are latent and abstract. They are always actualised by a multitude of *concrete* feelings and thoughts, which vary depending on the culture. Any culture might name these concrete feeling-thoughts in a different way.

Feeltypes are individual possibilities, but this factor does not lessen their importance. On the contrary, every psyche, whichever concrete form it takes (just like a body using five senses) has to use *all* of these possibilities. *This is the sole path to its existence.* Therefore, without understanding feeltypes, we cannot comprehend the reality of what we call the psyche. *We will remain lost among this multitude of affectivities.*

For all people past and present, the existence of the psyche is connected in some way or other to the experience of six fundamental feeltypes. As these feeltypes define humankind, *they are all crucial.* Everywhere the human species exists, every individual has a psyche, and every psyche is established with material gathered from the six axes. The absence of one is tantamount to *a disability*, just like a body bereft of an eye, leg, or arm. It is face-to-face relationships, shaped according to the surrounding culture of each person, that determine how these axes will be experienced in concrete terms. Doubtlessly, in the Stone Age, the nature of the "like" feeltype, for example, would have been substantially different to that of today. But regardless of its exact nature, any individual who cannot experience it in one way or another cannot completely fulfil their psychical being and functions.

In order to identify the feeltypes, it is imperative to accurately analyse both a specific culture and the concrete individuals that it is transmitted through. Just like the fact that we have two arms and two legs, this limited number of feeltypes in all individuals past and present is always *the same.* But as the feeltype source and the way they are experienced are societal, they *vary* according to location and time. For example, it is society that delineates what we will be angry at, why we will feel pleasure, and how we live both affectivities. In fact, the same logic applies to our body. Having two legs does not necessarily mean we are destined to crush grapes with our feet to make molasses, chase off enemies, or play football. Our societal relations determine what we will do with our legs. Therefore, the secret of what we will experience with our abstractly arranged feeltypes, and how, should be sought in the culture of every era and lived environment.

In the definition of the individual, the active determinants for the formation of the psyche are the feeltype's concrete origins and the patterns of their concrete formation. In other words, *the active determinant of individuals is their cultural interaction*, which emerges and is shaped by the face-to-face relationships of every individual juvenile. Alongside these, there are other factors – heredity for example – determining the individual psyche. But the traits acquired in this way are of *secondary* importance to the characteristics shaped culturally by face-to-face relationships and which fall along the six feeltype axes. These traits can be compared to bodily characteristics, such as being short, having black skin, long legs, or fair hair. They can only be *passive determinants* of the psyche. (Like all passive determinants, they can become active once they cross a certain quantitative threshold, but this has no bearing on the crux of the matter.) For example, whether a psyche that has been brutalised by growing up in a loveless environment is choleric or sanguine is of secondary importance, just as is the height of the body in which it lives.

In summary, the feeltypes theory is the most critical element of the theory of the psyche. This is because feeltypes: (a) define all affective-intellectual possibilities of humankind as a species, and consequently define all individuals past and present; (b) explain the variation in human nature across different cultures, because their existence is species-specific and their origin and form is cultural; (c) provide the tools for analysing cultures from a psychical respect; (d) allow us to understand how the psyche is established through the medium of culture; (e) provide the main vehicle for the theory of the psyche; (f) make it possible for us to concretely research the mediations that bring about the dialectical identity between individuals and society.

3. The Evolutionary Origin of Feeltypes

The only way to explain the *existence* of feeltypes is to look at evolution. Through evolution, we have developed a series of feelings that are powerful enough to seem as if they are part of us. They are all extremely functional, for instance, disgust and fear. Without disgust, there would be nothing to stop us satisfying our hunger with harmful or even poisonous substances. Without fear, we would not be able to fend off threats before they became deadly.

In societal relations, and consequently in the feeling-thoughts that originate from these relations, these somatic possibilities are frequently put to use. A type of social behaviour might disgust us, and the object of this disgust might be a person behaving in that way. What makes a particular behaviour disgusting is, of course, the fact that it conflicts with the values of the culture we belong to. However, the thing that enables us to experience this phenomenon in the form of a concrete feeling is the sense of disgust that has been passed down to us through evolution. The same applies to fear. It is our culture that defines what we find frightening in our societal relationships; but we experience it through the awakening of a sense of fear that we have inherited from evolution. Disgust and fear are the generic roots of the Like/Dislike and Trust/Fear feeltypes, and culture defines the reasons for their evocation. How they are experienced, and through which words and behaviours they are expressed are also dependant on culture. But the emotional tissue, the raw material of these feelings, is rooted in our species.

It is important to carefully distinguish between two types of somatic traits developed during the evolutionary process. On one hand, there is a set of needs (e.g. impulses, urges) associated with the natural reflexes that arise from our physiological build, such as breathing, motion, eating, copulation; on the other hand, there is a set of feelings that are functional to the life of the species. Fear and disgust have already been mentioned, but curiosity is another of such feelings: the curiosity to study the environment, survey the possibilities of the settlement area, and gather information useful to a particular purpose. Both types of phenomenon are biological and part of our genetic make-up, yet they are both still differential factors. The "feeling" of hunger and the feeling of curiosity are not qualitatively the same thing. The first stems from the need of a mechanism that must be satisfied in order for an individual and species to survive. The second is associated with a set of functional affective capabilities acquired by the body at a much more advanced stage of evolution. Feeltypes are related to these "more advanced" affective capabilities.

Sexual relationships are a useful example for helping us to understand the functional relationship of the culture-psyche-body network. It is culture that specifies who will show a sexual interest in who and which behaviour they will employ to show this. In the psyche, this is predominantly experienced on the Pleasure/Pain axis of the six basic feeltypes. As with all feeltypes, its roots are in the body, which has been determined by the developmental process of the species. Copulation is a natural reflex that was present even at the lowest stages of evolution. However, for beings living outside of a culture, like most animals, sexual intercourse is not a source of "pleasure"; it is a physiological practice just like filling our stomachs. Yet, at the highest stages of evolution, the natural sexual reflex can also be the means to a different kind of relationship. For example, bonobos derive pleasure from rubbing their genitals together or mutual stroking. This is not the same as standard sexual relations (copulation), which are much more biological. In humans, sexuality started as a purely natural reflex, but then began to take on different functions as its pleasurable aspect emerged during the evolutionary process, transforming into a cultural relationship in which the *pleasure aspect is the determinant.* Yet, if all these evolutionary stages had not unfolded, the *existence* of sexual pleasure, as seen in humans today, could not be explained. Culture only provides the object of an existing inclination and determines how it is experienced. It *does not create* biological inclinations, *it only transforms them.*

It may be a helpful analogy to think of the human psyche as a six-stringed instrument. Naturally, the qualities of the materials used to make each individual instrument differ to some degree, depending on the legacy left by the ontogenetic process. But ultimately they are all guitars, violins, or mandolins, etc. What decides whether they will be guitars, violins, or mandolins, in other words the determinant of the type/species, is phylogenesis – the product of evolution. Every string corresponds to a feeltype, while the melodies that will be played on the instrument are determined by culture. The format of each feeltype contains a multitude of tone possibilities (feeling-thoughts), from high- to low-pitched, from the most tremulous to the straightest note, from the harshest to the gentlest plucking of each string; added to this are the possibilities of mixed sounds (feeling-thoughts) resulting from playing more than one string at once. In the general determinant framework of their own culture, every individual acquires their own melody (character), which is a product of their own unique upbringing-experience. The psyche makes use of the body's possibilities when its own instrument is being formed and as it formulates its own melody. For example, like or disgust are dormant in the body. Using its Like/Dislike string, the psyche uses these dormant elements to produce tones (feelings-thoughts) that are appropriate to the format. A specific cultural space, along with the face-to-face interactions unique to particular persons within the space, enable a melody to be made from these notes.

As we can see from the feelings of repulsion, fear, and curiosity, the evolutionary physical sensibilities are already in place for *every one of* the six fundamental feeltypes. For example, physical pleasure is the basis for psychical pleasure. Comparable feelings of pleasure have developed in many animal species: as already mentioned, bonobos, especially the females, derive pleasure from rubbing their genitals together; chimpanzees pat each other's backs and pick out their fleas, while dogs and cats "enjoy" being stroked on the head, back, and stomach by humans; numerous animal species lick each other, etc. Another example is that at the heart of the psychical Like/Dislike feeltype is an evolutionary feeling of affinity or hostility. Many animal species feel an "affinity" towards others, especially towards juveniles and sometimes towards mates/counterparts or members of the group sharing the same area of land, as they display "anger" towards sexual competitors, food rivals, and those from outside the group.

For another example, the Approval/Disapproval feeltype is based on the feeling of "respect", or its opposite, which is imprinted in the genes by evolution. In some advanced ape species that live communally, there is a very effective hierarchy to determine the choice of mate and apportioning of food. Hence, it should come as no surprise that the feeling of "recognition" of one's status within one's community has evolved as a species-defining characteristic. Equally, underlying the psychical Happiness/Sadness feeltype is a feeling of "satisfaction" observed in advanced animal species, particularly dogs and cats. Many other animal species display this feeling when faced with a favourite food, or some animals in zoos react to their captivity by becoming "melancholy". When the physical sensibilities of repulsion, fear, and curiosity (Comprehension/Confusion feeltype) are added to those I have listed, it is not hard to see that an evolutionary sensibility exists, upon which the six basic feeltypes are based *without exception*.

However, it would be a great *mistake* to jump from this point to the belief that it explains the origin of psychical characteristics. The individual psyche *originates* from society. *Both the content and mode of experience of all the feeltypes is provided by society (culture)*. That said, it is not a complete irrelevance that we have a human body rather than that of, for instance, a crocodile. While the body may be a passive determinant, it is nevertheless a determinant. Our body, with its form, strength, senses, etc., gives both limitations and possibilities. And among these possibilities are also the formations of feeling-thoughts accumulated in the body. It is crucially important to grasp this. Just like the ability to walk on two legs, these too are a set of *abilities* arrived at through evolution and imprinted into the generic genetic structure of humans. These abilities provide the building materials of the psyches that emerge in individual humans, the only social species. For a baby emerging from its mother's womb to "recognise" the feeling-thoughts encountered in face-to-face relationships right from its first day, transforming them into the matter of its own psyche through an active process of interaction, can only be explained by these feeling-thought types being *dormant* in the body. Just as a baby engages in inter-individual interactions with five senses – seeing, hearing, smelling, touching, and tasting – it accesses the feeling-thought dimension of this same network with the six basic feeltypes already in the body – the *capabilities* of affection, trust, approval, joy, pleasure, curiosity. A child could not hear if it was not for the capability known as hearing already being part of the body. Likewise, a child could not have the feeling-thoughts of fear, like, or curiosity if these capabilities were not latent in the body. Of course, what the child hears depends on their environment (society, culture), as does the subject of their likes, fears, or contemplations. Whether it is a sensation or feeling-thought, all these *capabilities* are already woven into our genetic structure by evolution. Therefore, if the body is able to convey the psyche, the secret must be that the body has previously undergone some formatting to optimise it for the purpose.

As part of Darwin's efforts to prove that humankind is also a product of evolution, he repeatedly emphasises the animal origins of human feeling-thought in the *Descent of Man*. He argues that even the most distinguishing characteristic of humans, their mental powers, can be explained by evolution: "Nevertheless the difference between man and the higher animals, great as it is, is certainly one of degree and not of kind. We have seen that the senses and intuitions, the various emotions and faculties, such as love, memory, attention, curiosity, imitation, reason, etc., of which man boasts, may be found in an incipient, or even sometimes in a well-developed condition, in the lower animals."[4] However, Darwin overlooks the transformation of quantity into quality during the evolutionary process, writing: "But everyone who admits the general principle of evolution must see that the mental powers of the higher animals, which are the same in kind with those of mankind, though so different in degree, are capable of advancement."[5] In fact, Darwin was confusing two different phenomena. It is true that in order for a person to experience a feeling, they must have acquired the relevant sensibility through evolution. Evolution is the *root* of the feeling-thought capability. However, its *source* is society rather than evolution. The interesting thing is that when analysing a concrete feeling, such as cooperation, Darwin argues that it is connected to societal relations rather than evolution: "The motive to give aid is likewise somewhat modified in man: it no longer consists solely of a blind instinctive impulse, but is largely influenced by the praise or blame of his fellow men."[6] As this example clearly shows, when it comes to humankind, the *real source* of feeling-thoughts is relationships with others. Just as physiological evolution gives humans the potential for bipedalism, speech, and using hands to perform certain tasks, it also gives them a series of affective-intellectual *abilities*. But what equips these abilities, reshaping them according to the equipment, enabling them to be used properly, and therefore providing their real source, is culture. For example, it is community relations that dictate which societal phenomena trigger the desire to give aid, which appeals and expectations will be responded to, how aid will be delivered, and which mental configurations it will forge in the individual.

In fact, the argument put forward in this book in relation to the social nature of the human psyche is entirely in keeping with Darwin's understanding of evolution. In Darwinist evolution, a higher stage is not the "result" of the one before. If that was the case, there would be a "goal", an evolutionary direction determined at its outset, which the chain of results would gravitate towards. But there is no such thing. The direction of evolution at every stage is determined by the conditions *of that stage*. However, in order to respond to the necessities of that stage, the relevant species *uses* the possibilities/capabilities made available by the previous stage. Those who use these capabilities most successfully sit at the top of the natural selection order. As society was established and social interactions were constructed, the being that was to develop into a human formed the psychical elements necessary for these relationships, using the set of affective-intellectual capabilities that were already present at the stage they had reached. However, when these capabilities were subsequently used for *societal* rather than zoological relations, their *qualities changed*. Zoological sensibilities turn to human-specific sensibilities. Obviously, in order for this transformation to take place, these sensibilities must be in *existence*, but the necessity of their existence *does not make them into a "cause"*. The cause is society. And the result is social: humankind as a unique social species.

In short, the feeltypes that make it possible to lay the building blocks of the psyche are rooted in evolution, although their content and shape are determined by social life. Like everything else, they emerged through a qualitative change in *something that already existed*. The element that underwent qualitative change to take on the form of feeltypes in humans was a specific set of *sensibility capabilities*, which emerged through evolution and were imprinted into the genetic structure as a result of natural selection.

4. The Formation of the Soul

Human juveniles enter the world as an organism carrying the genealogy (phylogenesis) of their species, their own genetic build (ontogenesis), and the conditioning experienced in the mother's womb. Every child is born with a certain genetic potential, and a certain nerve cell (neuron) and hormone potential. Conditioning arises from the various influences experienced by the embryo and the foetus during gestation, beginning with the first fertilised cell. As soon as the child is born, as well as its existing unconditioned (natural) reflexes, it begins to acquire conditioned reflexes. The human juvenile is subjected to a ceaseless interactive flow through their five senses. Every effect that is experienced becomes the source of a conditioned reflex. However, these reflexes that are acquired even before birth are comparatively unimportant with regard to the development of the character. The active determinant in the formation of the personality are the effects of the social environment into which a child is born.

The marks left by the social environment form the first building blocks of a child's psyche by constantly reshaping its natural biological state. As these reflexes accumulate, the child's psyche is configured. Conditioned reflexes are none other than individual habits, and these habits establish the personality as an unconscious system. As they become increasingly recognised, the conscious part of the psychical system begins to form. However, the unconscious and conscious system are in no way separate from each other. They are intertwined. The unconscious and conscious areas are malleable and transitional, possessing the capacity to transform into each other.

Perceptions gleaned from the social environment are initially in wholesale form. This applies not just to the sense of sight, which takes in the outside like a camera, but to *all the senses.* At this stage, there is no consciousness. Up until four or five months of age, a child cannot differentiate between itself and its environment. The perception of itself as a separate entity begins afterwards: "We refer to the psychological birth of the individual as the *separation-individuation process...* Like all intrapsychic processes, this one reverberates throughout the life cycle; it is never finished... But the principal psychological achievements of this process take place in the period from about the fourth or fifth month to the thirtieth or thirty-sixth month, a period we refer to as the *separation-individuation phase.*"[7] During this period, the infant surmounts the two most important steps of humanisation; at around one year it starts to walk and at around two years it starts to talk.

Like all the other acquisitions, language is also acquired via conditioned reflexes. Comprehension of certain human sounds, not just a look or a movement, as the signs for other aspects of reality is of critical importance in the humanisation process. As these signs accumulate, the wholesale photographs created by the senses slowly become individualised. This separation occurs due to societal stimuli and conditioning. Language itself begins like a photographic perception. Consciousness begins to develop as the photographs become pliable and separate into details during the process of social interaction, making these detailed perceptions part of a general perception.

A child uses a photographic language to familiarise and communicate with its environment. At around school age, this language gradually turns into an inner language.[8] The transformation is stimulated by society. Using this language, a child produces thoughts and feelings related to itself and its environment. Initially, this process is confined to concrete phenomena, as the child's signs and symbols have not yet developed a sufficient abstractness. Getting to grips with abstraction is a long process, and it is only on reaching adolescence that the individual learns to think abstractly and use concepts as signs.

Throughout this entire humanisation process, the active determinant is social relations. In order to acquire the necessary framework for these relations, the organism needs to undergo a certain development process, which is organic and biological. However, it is crucial to understand that it is always social interaction that catalyses and sustains this organic development. Of course, a child must undergo a certain process of organic maturity before being able to walk and talk. But without social exchange, it would not be able to learn either skills. Therefore, biology is only a *passive* determinant. The *active determinant* is the *social interaction* that takes place within each specific culture.

What is important for the healthy development of a human juvenile is that the basic positive feeltypes (the first element of each axis) receive sufficient stimulation and satisfaction from the body-to-body relations which begin at birth. The dominance of a basic negative feeltype (the last element of every axis) injures or handicaps that region of the psyche. As already discussed, it is not just the source of feelypes that is social, but also their forms. For instance, society determines who will be loved (some types of affection are forbidden, some are encouraged) and how (to what degree, in what form, through which means, the manner of togetherness). The object of love or fear is not abstract, it is an extremely concrete person or persons (with a specific sex, hair colour, voice tone, gait, look, smell, embrace, etc.), who play a specific role and possess certain physical-psychical characteristics.

Through face-to-face relationships, the psyche forms a template that is, to varying degrees, suitable or unsuitable for emptying and replenishing its energy. The more homogenised conduct becomes, the more it is consolidated. This template is largely subconscious. The specific colours of the template are painted by the concrete face-to-face relationships of the individual, and the psyche is built from the concrete data arising from these relationships. It is impossible for a child to feel abstract love, distress, etc. It is always a relationship with a *certain person* that causes distress, provokes love, or incites fear.

While it is true that the source of such psychical assignations can also be objects (e.g. natural events, light, noise, sunrise, the fragrance of a flower, dogs, birds), these assignations are of secondary importance to those arising from inter-individual relations, because they are in reality derivatives. *As people acquire* their human sensitivities from face-to-face relationships, they *gain the ability* to give human reactions when interacting with objects.

Social interaction, which starts even before birth, incrementally builds the individual psyche. The psyche is entirely subconscious at the beginning and continues to be so to a large extent. A condition of the psyche's existence is that it is able to maintain a degree of inner balance as a system and have a more or less regular energy exchange with its environment. The psyche is formed within the framework of these limitations and necessities. In reality, this process of psychical formation carries on until death. But as is inevitable when creating a system, the early years of formation are *rapid*, because the psyche can only exist at all in the form of a system or structure. Systemisation and structuring are *conditions of the psyche's existence*. A disorderly mass of feeling-thought would not function as a psyche. Therefore, rapid progression into a system is a matter of life and death for the psyche. To be able to be human, the human juvenile must complete the process of developing a psyche, or at the very least a rough draft of it, as quickly as possible. Inevitably, this to a large extent places conditions on any later formations. The psychical system acquired in the early years to some degree determines the individual throughout their life.

Feeltypes can only be experienced *concretely*, and therefore every individual experiences them uniquely, in a *relatively different* way to all other individuals. The common feeltypes, which occur in the same assortment and number in all individuals, just like eyes, arms, and legs, have *relatively different* content in every individual. What differentiates this content within the same culture is the unique individual life of every person, the specificity of those with whom social relationships are formed and the different degree of the influences experienced in different relationships.

As psychical discrepancies accumulate, hindering the replenishment of energy, the individual constantly develops balancing reactions in response (Freud's defence mechanisms). Over time, these mechanisms become fixed, and consequently, as described by Adler, the unique unchangeable game emerges, which every individual plays lifelong;[9] in order to renew their energy, the individual keeps repeating their personal customised versions of these games, whose main themes are sympathy, deceit, aggression, dishonesty, etc. This fundamental game leaves its mark on the character of the person concerned.

The traces left by psychical experiences vary according to the intensity of emotions provoked when they occur. Some experiences are such that they are remembered for a lifetime. The reason they are not forgotten is connected more to the permanent transformations they cause to the psychical balance than to the depth of the traces they leave behind. These transformations precipitate powerful currents, like a vortex, repeatedly pulling later experiences towards the same foci. The individual feels-evaluates everything they experience in terms of its effect on these foci.

Experiences that leave deep traces are mostly negative ones, occurring at the negative extremes of the six main feeltypes. A person may feel completely unloved, drowned in despair, totally humiliated, wracked with fear, crippled with pain, or numb with shock; these are emotions that usually occur suddenly, while their positive counterparts normally occur over a prolonged period of time. Nonetheless, there are anomalous sudden and intensive positive examples of each feeltype; a person can fall madly in love, feel deeply relaxed by a feeling of total trust, be aroused by the excitement of attraction to others, be overcome by happiness, want to jump for joy, be elated by physical pleasure, or feel exhilarated by learning or discovery.[10] In these circumstances, the experiences have a permanently transformative effect on the psychical balance. As a general rule, loss of balance occurs suddenly but takes time to recover, and the same is true of psychical balance. Contrary to their negative counterparts, constructive positive experiences, like comforting, empowering, or progressing, are not sudden. This suddenness is a core quality in the nature of negativity. Suddenness is, by definition, at the opposite end of the spectrum to the concept of balance.

Any sudden negative experience happening to an individual takes place in the body, just like breaking a bone or pulling a muscle. Any of the nerve networks in motion when the evoked emotion is experienced become injured. But in comparison to the relatively quick recovery period of physical injuries, psychical damage is difficult to heal. The reason is easy to understand: physical injuries heal through natural physiological processes. They can be facilitated or accelerated through conscious intervention, but the main recovery factor is still physiological. On the other hand, psychical injuries originate from society. An individual is psychically impaired by being in a certain social environment. Once this impairment has initially occurred, it will continue to define the individual's relations with their environment. As a result, the environment has a generally destructive rather than ameliorative effect. While it is also true that physical impairments might determine a person's social relations, this would signify that the physical impairments have ceased to be purely physical and turned into psychical impairments. Suffering due to disabilities is often due to mental rather than bodily pain. In other words, it is psychical.

These wounds and impairments always impact on the consequent life of the individual. In response to the anxiety and depression they create, the individual develops defence mechanisms. Conscious mechanisms are not problematic. On the contrary, they help to re-establish the psychical balance. Yet, subconscious mechanisms, literally defence mechanisms, become breeding grounds for further mental suffering. They drag the individual towards pathogenic defence behaviours, making their social relations increasingly artificial and gradually making the chances of recovery more remote.

The traces left by these negative or anomalously positive experiences are not immutable. On the contrary, subsequent experiences continually affect them. A child who experiences intense fear, and in later life does not have a sufficiently secure environment, gradually becomes more insecure, withdrawn, and fearful, still remembering the hugely negative initial source of fear with dread and distress. In contrast, a child who has the chance to continue their life in a more secure environment gradually develops indifference towards the previous frightening episode, perhaps even remembering it with a smile. Therefore, our psyche determines the feelings that are evoked by the memory of a past event that has marked us, determining the psychical state we are in for the rest of our life, particularly when we recall the event.

The recall of exceptionally positive experiences is subject to the same rule. For an aged person who is no longer interested in sex, the memory of an ecstatic passionate experience is no longer the source of arousal; at best it creates a feeling that provokes a smile. When someone who has grown up in a safe environment is rescued from a situation of intense fear, the feeling of immense trust that ensues leads to a feeling of cheering, pleasurable warmth rather than a new source of security. When someone thinks back to a moment of great realisation, when things fell into place or the curtain was lifted from their eyes, it doubtlessly evokes a feeling of pleasure, but this is no longer the excitement of the initial comprehension. This was lived once and has now passed. Only a new discovery can trigger this level of feeling.

In short, everything we experience leaves its mark on our psyche, shaping it and tipping the balance in varying degrees. When we recall any event, we experience it only in the context of all these past developments and according to the characteristics of our present psychical framework. Just as feeling and thought cannot be separated in the psyche, the act of recall can never be a simple act of visualising a chunk of knowledge. This visualisation always stimulates some feelings, which also change according to the nature of the psychical configuration. Sometimes this change can be so powerful that it even changes the actual knowledge of the event, turning it into something else. A person may subconsciously disown their past life, changing it and even to some extent making it up.

The ratio of unconscious to conscious in the psyche is a product of the life and culture that has shaped the individual. Cultures that provide different, contradictory determinations stimulate the expansion of individual consciousnesses. The same is true for individual lives built on varied experiences. The greater the diversity and richness of our social relationships, the greater is our consciousness. However, it would be impossible to have an entirely conscious psychical framework. No psychical framework could exist without the help of unconscious conditioned and unconditioned reflexes and habits; it would be like trying to live without any sleep. On the other hand, it is possible to live with a largely unconscious psyche, depending on the immaturity of the culture or an individual's childhood.

What humanises humans is their ability to consciously make significant changes to their environment and, consequently, to themselves. The most powerful medium for achieving this is consciousness, and therefore consciousness is considered tantamount to humanisation. The richer the social relations of a society are, or in other words, the more conscious are the psyches that form in it, the more possible it is for people to raise the level of humanity in that society. *It is consciousness that makes humanity.*

5. *The Structure of the Soul*

For the purposes of analysing the soul's structure, the topographical model first developed by Freud, in which he classifies states of the psyche as unconscious, pre-conscious, and conscious is clearer than the structural model he later developed. Freud's three-tier structural model is problematic. According to his model, the psyche is an affective-intellectual system consisting of id, ego, and superego. While Freud's analysis of ego and superego is useful, his analysis of id is considerably problematic. The root of the problem lies once again with Freud's connection of id with impulses, in particular sexual ones. There is no such connection. *All parts* of the psyche, without exception, originate from inter-individual relationships. The structural model can only be usefully employed if this *basic reality* is borne in mind.

One of Freud's greatest contributions was his discovery that the psyche was largely unconscious. It is true that the id is unconscious and determines a significant part of our behaviour. It has many components, which can be grouped in two large clusters. To begin with, all our unconscious *habits* are here. The feeling-thought habits that we have accumulated on the six basic feeltype axes are imprinted on our id, just as when we learn to ride a bicycle the coordinated motion patterns that enable us to keep our balance are forged into our body. This is the first large cluster. However, what makes the id interesting in psychology is that the unconscious *defence mechanisms* we develop to deal with problems we encounter in our dealings around basic feeltypes also accumulate here. This is the second big cluster. Let us suppose that we find it practical to resort to violence or introversion in response to negative events we experience on the Like/Dislike axis. This reaction forms a feeling-thought pattern through constant repetition, shaping our id accordingly. Over time, as we have similar reactions to similar circumstances, this pattern of reaction turns into an unconscious attribute of our character.

A comprehensive understanding of *defence mechanisms* is a prerequisite to appreciating the workings of the psyche. The purpose of these mechanisms is to conceal inclinations that give rise to anxiety or psychical pain and to bring them under some degree of control. There are numerous detailed explanations of this in the annals of psychoanalysis, and it might be useful to list some of them here. The individual represses or denies a discomforting inclination in the following ways: by associating it with others but not themselves; by acquiring a behaviour pattern that is the complete opposite; by trying to legitimise the inclination on rational grounds; by immersing themselves in intellectual research about it to discharge some of the affective load; by trying to distract attention from it with obsessive behaviours; by blacking out if they cannot bear the mental pressure; if they cannot blackout, by inflicting physical self-pain that is strong enough to suppress the psychical pain; by turning the object of the inclination in a less dangerous direction or towards themselves; by suppressing the painful inclination with lofty deeds, etc., if they have the means to do so.

The unconsciousness of the id originates from the fact that the baby in its first stages has still not reached the level of discernment. By staying out of the conscious realm, the psychical habits acquired during this period fulfil their functions without unnecessarily burdening consciousness. Then, as the consciousness develops, the psyche expels from it the relationship patterns that it realises generate pain, fear, and anxiety. Thus, the feeling-thought patterns that the consciousness *wishes to avoid* begin to build up in the id *alongside* the feeling-thought patterns that accumulate automatically. Features acquired earlier and later in life intermingle, reciprocally remoulding each other. This is how the steps of the id's development come together.

Like the id, the superego is also unconscious. It is the pattern of societal norms woven inside us, the internalisation of norms, along the psyche's six basic axes, which determines what is right and wrong, good or bad, beautiful or ugly according to the relevant culture. Superego and id are unconscious. However, while superego forms from the internalisation of societal-cultural norms, the id forms from the marks left by an individual's positive or negative experiences. Positivity and negativity are defined according to the extremes of the six axes. The human juvenile tries to repeat positive experiences and avoid the negative ones. When necessary, this avoidance occurs through simple or complex defence mechanisms that obscure the avoidance itself. Over time, the superego and ego install components in the id. The superego, just like the first layers of the id, unconsciously acquires a number of habits. If these clash with some of the id's feeling-thought patterns, the id acquires a characteristic that is contradictory and conflicting in the relevant respect.

It is worth repeating that the actual contents of the six axes and the nature of the positive and negative extremes, etc., are all *cultural.* Contrary to the thoughts of Freud and others carried away by physiological determinism, there is no immutable series of physiological tendencies that defines them. The countless cultural norms that clearly conflict with physiological processes are clear evidence of this.

Along with the id and superego, the third part of the psyche, the ego, is also initially completely unconscious. We know that bipedalism is one generic characteristic of the human species. This is a significant phase during the formation process of the psyche and has a universal nature.

However, the fundamental threshold is the formation of the feeling of: "I am a separate entity in the environment." The emergence of this feeling of ego, in the broadest sense, for every individual is a universal phenomenon. As already stated, the content of this ego is at the beginning entirely *unconscious*. The process of intellectual development referred to here was essentially propounded by Piaget. As he showed, every human juvenile, regardless of which culture they belong to, undergoes a process of "recognising themselves as themselves". At some point during the growth process, every juvenile begins to perceive the separation of "you" and "me".[11]

This sense of self, and of belonging to a species, can only exist in a relationship with others. The sense of "I am a separate entity", which triggers consciousness, is identical to the sense of "I am part of you", because humans can only be a member of the species as individuals. It is only when a human perceives themselves as an integral part of society that they perceive themselves as a separate "me".[12]

Just like the body's fight for survival, every individual needs to develop this sense of self, or ego, in order to maintain their human quality. Consequently, having and developing relationships with others is essential for the life of every psyche. The relationship is vital. Even the most powerful somatic needs can only be met in the manner and degree allowed by this relationship. For example, a child eats to reward those they love or shuns food to punish them, even without being consciously aware of it. Depending on the behaviours of those in its environment, a child may or may not wet themselves, may cry or laugh, sleep or not sleep.

As pointed out, initially these childhood behaviours are unconsciously determined. But the culture of the society decides when the child will reach an age of consciousness or the level of consciousness attainable in the society concerned. It is also the case that reaching consciousness always necessitates a certain physio-psychical development process. The rule is that the psyche is unconscious. Therefore, when the individual is undergoing this development process, their compliance with the need to preserve their individuality is initially an unconscious mechanism. This unconscious mechanism also casts every psyche in a specific mould. This is the primary source of individual character.

Whatever age we are, the main determinant of our behaviour is the id. Just as taking a step does not require us to consciously recall the rules of walking, our behaviours also unconsciously comply with psychical habits. While some of these are as unproblematic as walking, others are problematic to varying degrees. In the process of the inter-individual relationships we experience along the six main feeltype axes, anxiety is created by behaviours that arise from unconsciously developed patterns of reaction (defence mechanisms) to societal rules (the source of superego) or from the improper behaviour of others. Let us take as an example a gratification we have contemplated or enjoyed in our inter-individual relationships, which, due to a feeling of shame instilled in us by society, in some way conflicts with the rules we have absorbed. Whenever we contemplate this gratification or have the opportunity to experience it, the societal prohibition imprinted on us creates an inner turmoil, regardless of whether we actually act on our desire or indulge in dreams about it. In the same way, unconscious defence mechanisms that have developed towards previously experienced negative behaviours click into action whenever we encounter comparable negativities, or even when they come to mind or if there is a possibility that they will occur. This also causes anxiety.

Anxiety is a fear whose origins we are not aware of. We can only discover its origin by expanding our field of consciousness. However, if our ego is so frail that it cannot bear reality, consciousness can further exacerbate our psychical pain, illness, wound, or impairment. Therefore, lack of consciousness is not a structural flaw. On the contrary, it is mostly a *functional* issue. On one hand, as just pointed out, it serves to obscure some unbearable realities. On the other hand, it fulfils a condition of existence for psychical life. If not for these unconscious habits, i.e. if we had to consciously think about how to take every step, life would be impossible.

We encounter the same phenomenon in our inter-individual relationships. Much of the time, unconsciousness, i.e. the concealment of some realities either wittingly or unwittingly, is a necessity for the maintenance of a relationship. Hence, the common platitude of, “Your relationship will be better if you are open about everything”, is generally somewhat unrealistic. Enhanced consciousness, or everything being out in the open, may well destroy rather than mend a relationship. The mutual unmasking of defence mechanisms may even incapacitate the very pathogenic accord they effect. If we try to cast the searchlight of consciousness over every habit, these habits could lose their ability to function as mainstays of the relationship.

In a society full of contradictions, it is inevitable that the individual id is also full of contradictions. Contradictions, both in and between societal roles (or in other words structural contradictions) and contradictions arising from the clash between the characteristics of each individual and societal tendencies, make the id a mechanism founded on dissonant feeling-thoughts, all loaded with organismic energy, which intersect at some point of the basic feeltypes. It should not surprise us that whenever these mechanisms operate, they create anxiety and psychical pain. The individual (and consequently limited) solution is to strengthen the ego with consciousness and conscious behaviours. The radical solution is to establish a new society, one which creates id mechanisms that do not provoke such degrees of anxiety: in other words, a revolution.

As implicit in the name, the psyche is psychical. It is located in biology, yet it is not biological of itself. It is a feeling-thought combination. The system we refer to as the psyche requires that all its axes function positively and cohesively. Positivity correlates with habits close to ends of the axes that gratify the psyche. Cohesion originates from the psyche's systemic quality. Every system requires a structure and mechanism that keeps the axes from clashing; and while positive habits help to achieve this, defence mechanisms undermine it.

As already mentioned, from birth onwards, like every other organism, the human juvenile possesses a sequence of reflexes and requirements (inclinations, impulses) that enable it to stay alive and develop. The qualities of these reactions and needs is determined by the societal medium through which they are gratified. A baby breastfed by its mother will feel the need for her breast. If the baby is fed by a wet nurse, it will feel the need for her breast, and if it is fed with a bottle it will feel the need for the bottle and the warm hand that holds it. When these socially conditioned needs meet with obstacles and negativities, the baby begins an unconscious quest, developing resistances and introversions. A child's gratification or deprivation are determined by the inter-individual relationships in which they are involved. The energy for feeling-thoughts and motion undoubtedly comes from the body. The sense of gratification or deprivation is also experienced in the body. But it is inter-individual relationships that give it form and colour. Therefore, it would be misleading to take a fundamental need like breathing as a basis for explanation of how the psyche is created. It is equally wrong to take other fundamental needs like eating or sexual satisfaction as a basis. Specific *obstacles* stemming from the child's environment make a particular inclination *problematic*. For instance, if the child is kept in conditions where it is difficult to breathe, it will develop reflexes and defences in that area. The defining factor is not the inclination itself, but rather the conditioning arising from the outside environment, which delivers varying degrees of satisfaction or inhibition. The baby's id is shaped according to these influences.

The term "defence mechanisms" was coined by Freud and his followers. They brilliantly defined most of the workings of these unconscious general behaviour templates, which try to protect malformed or pathogenic psychical systems. Which dangers does the individual need to protect themselves against with these defence mechanisms? In my view, the threats to an individual originate from two sources: (a) being *excluded* from the social environment in which they experience face-to-face relationships; (b) being so *engulfed* by this environment that they cannot individualise.

The earliest and greatest threat an individual might encounter during their developmental stages is the first of these, because an individual alienated from these relations *cannot humanise*. This even threatens the individual's physiological existence and development. Therefore, when the individual psyche is forming, it is first and foremost forced to develop mechanisms of conformance with those around it. These mechanisms of conformance may include elements of aggression. If an individual perceives threats arising from their environment, they will undoubtedly develop defence mechanisms that contain aggression. The dominant values of each culture define which elements of a child's behaviour are liked and disliked. Being liked, assimilated, and receiving approval from those around are conditions of existence for the ego. What are referred to as the defence mechanisms of the ego are actually the basic behaviour patterns formed subconsciously by each child in order to fulfil this condition of existence. During its formative stages, the psyche is shaped by these templates.

Freud researched how these templates form within his own cultural environment, resulting in some important discoveries. However, Freud made two major errors: (a) he saw these templates as common to all people past and present; (b) he connected the formation of these templates to unchanging physical impulses that he saw as unique to the human race. Once Freud's findings are purged of these errors, there is much to be learnt from them. Furthermore, as Freud's cultural environment was little different to our own, the degree to which these findings are valid for our own environment is much greater than for completely different cultures.

Some of these mechanisms can also be aimed at eliminating shortcomings that the individual perceives they have when seeing themselves through the eyes of others. For the individual, these shortcomings are the source of a danger that may prevent them occupying their desired position within their face-to-face interactions. This explains a defence mechanism that Adler attached great importance to: compensation aggression. An individual who sees themselves as inferior to others, for example with respect to height, arm strength, or speech, will go to extraordinary lengths to hide whichever shortcoming they think is preventing them from attaining their desired status among others.[13]

In that case, defence mechanisms primarily originate from outside threats that can hinder complete social fusion and from the personal shortcomings perceived by the individual as threats to this fusion. These are the main causes of defence mechanisms in all humans.

However, humans also have a need to individualise to a level parallel with the progress reached by their society. The formation and development of individual consciousness progressively increases this need, once more presenting a twofold threat to the ego: firstly, a suffocating cluster of relations that leave no room for individualisation, and secondly, personal shortcomings of a nature that will hinder individualisation. Adler's "compensatory reactions" originate from here. The individual declares war on the personal characteristics they perceive as deficient to the ideal self-image they have constructed in their mind, and every behaviour is determined by this compensatory quest.

As already outlined, the prime need of the psyche is to establish positive inter-psyche relations and the second is to individualise, especially in societies that have reached a certain level of development. The greatest threat to the first, basic need is social exclusion, which poses a lethal threat to the psyche. The greatest threat to the second subordinate need is being suffocated by social interactions. If the primary threat cannot be congruously warded off, the individual develops defence mechanisms. These mechanisms cannot be applied to the second threat, because defence mechanisms are always unconscious, whereas individualisation is always a conscious effort. In fact, these defence mechanisms can only ever play an obstructive role in the process of individualisation. Moreover, individualisation occurs through strengthening the ego and overcoming these mechanisms through consciousness.

Looked at from this point of view, how can we explain Freud's two main drives, Eros and Thanatos (Life and Death instincts)? When describing these drives,[14] Freud made some crucial observations about human psychological reality. His fundamental mistake was to ascribe physiological roots to these drives, whereas, as I argue, both these drives originate from society. The individual's greatest source of joy is to *harness inter-individual relationships to shape* and satisfy the pleasures *enabled* by the body. For example, satisfying the need for fresh air by enjoying the countryside with friends, satisfying the urge for movement by playing with friends or participating in team sports, satisfying the need to feed ourselves by eating together with friends, satisfying the prospects of sexual arousal through sexual relationships with those we are attracted to.

The mere thought of these gratifications can give the individual pleasure, and therefore the artistic activities that reflect them provoke feelings of pleasure and accomplishment. These are the most human characteristics. It is not the body that is involved in these relations, but rather the psyche. It is a complete mistake to see two people dining or having sex together as just individual bodies. *It is a biologistic deviation.* The bodies in such a situation are bodies already shaped by a psyche imprinted into them (in their synapses, hormones, reflexes, stance, looks, habits, dress, jewellery, the perfume they wear, the way they move). The copulation of two bears on a mountain is a physiological occurrence. The copulation of two people in bed is a societal, or in other words, a *psychical* occurrence.

If we understand Freud's Eros (life instinct) purely in this way, we can find an explanation of scientific value. Yes, Eros is the individual's most powerful life activity. It always exists in a web of social interactions. The bodies that facilitate it, with their innate energy and capability for pleasure and gratification, are just a vehicle for the psyche. In reality, as far as the human species is concerned, *there is no such thing* as a body that has a purely physiological quality. The human body is always a body imprinted with the mark of the psyche. What we see when we look at ourselves or others are not individual physiological objects, but individual people, or in other words, the individual form of society.

Freud drove himself into a dead end trying to understand the psychical drive he called Thanatos (death instinct) on a physiological plane. This explains why later psychoanalysts chose to overlook the drive. Nevertheless, just as with his observations of Eros, when investigating the idea of Thanatos, Freud made some scientific observations. It is true that in some situations, human behaviour displays death-kill tendencies. So, how can these be explained?

Freud sought the answer to this question in biology. He hypothesised that organic matter harboured a tendency to return to the non-organic elements it was made up of. He attempted to explain the individual death quest with this tendency of organic matter. Naturally, he failed, because trying to explain behaviour with physiology leads into a blind alley.

Thanatos is a mechanism of attack against the outside environment and the self. Like all defence mechanisms, the individual develops this in two basic situations: either they have been excluded from the outside environment or they are being stifled by the outside environment. Both situations pose a mortal threat to the individual, so the individual develops kill-die reactions, becoming aggressive towards the outside world or themselves. However, in the first situation they use unconscious mechanisms, and in the second they use conscious mechanisms.

Nevertheless, just like Eros, Thanatos may be an effective scientific concept for explaining human behaviour, but only if it is based on the social rather than the biological. If a betrayed person kills their lover or themselves, it is not related to biology. A body that desires to be erased from life is simply the carrier of a psyche that is the source of pain. People do not kill themselves because their cells have rejected organicism. Suicide means being unable to bear the pain of the psyche anymore. What else can be done when the only way to kill the psyche is to destroy the body it is imprinted on and of which it is an inseparable part? Ultimately, Thanatos is an entirely societal phenomenon. It is a defence mechanism employed against pain and threats arising from the social interactions inside which the psyche, the individual form of society, exists; it is a form of self-defence that uses aggression as a medium.

Every individual, healthy or unhealthy, is forced to maintain some degree of psychical integrity, inner balance, and energy exchange with their environment. This is achieved by acquiring a certain pattern of behaviour. This pattern, whether healthy or unhealthy, is unconscious in every individual. When the psyche is impaired or ill, this pattern of behaviour cannot quite fulfil its functions. It is only partially successful in maintaining inner balance and enabling energy exchange. An individual weighed down by the pressure of either insufficient or excessive accumulation of energy lives in constant distress for reasons they are unaware of. The distress may vary in intensity from person to person, but it is always there.

As explained, the body's exchange of energy with society is always a psychical phenomenon. A defence mechanism is an exchange pattern that executes this societal energy exchange with prodigal extravagance at great cost to both society and the individual. The only way to end this profligacy, anxiety, and pain is to break the pattern. However, the pattern is unconscious. So, the only way to overcome it is to nurture awareness of the defence mechanism and replace it with a new pattern of behaviour that will actually maintain inner balance and execute the energy exchange more prudently. Consequently, curing the psyche is tantamount to heightening consciousness, and therefore, the ego. The only way to provide any degree of relief for the tangled mass of dissatisfactions, reactions, contradictions, defence mechanisms, etc., that constitute the id is to try and bring it into the consciousness, particularly its painful aspects, and resolve it by acquiring new patterns of behaviour through the conscious behaviours of the ego. In other words, the ego must be strengthened. When these new patterns of behaviour become habitual and unconscious, they are established in the id, where they can then support its balance and health. Hence, they help to make the individual's psychical life relatively less painful and more enjoyable.

6. The Location of the Soul

The location of the psyche is clear. The psyche is imprinted into the entire body, from our neurons to our posture. Belief in the existence of a psyche outside of the body is completely unscientific. The psyche and the body are connected in every way. They come into existence together, and the death of the body is also the death of the psyche. The mediation of all characteristics of the psyche is, without exception, organic. Enis Batur's book, *The Body*, was published in 2007. The issue that Batur endeavoured to answer with this culmination of almost seven years' work was the bodily manifestation of the psyche. As every psyche is unique, so is each body.[15] People do not just talk with their mouths, they also talk with their hands and feet.[16] The psyche is so woven into the body and shapes it to such an extent that even its most typical characteristics are verbally expressed through physical idioms. It should suffice to list just a few of the many examples listed by Batur with his artistic sensibility: big head, eyes bigger than your belly, grasping, no stomach for it, heartless, no guts, big mouth, etc.[17] All these physical idioms really describe a psychical characteristic. Would anyone deny that the eyes are ultimately the windows of the soul? Even if the eyes are always part of the same bodily form, their looks differ. Batur listed forty-three examples of different looks from pensive to blank, calm to crazed. Importantly, his list ends in an ellipsis to stress that the list is open-ended.[18]

According to Freud, the psyche is a system with its own energy. It is true that the psyche is a separate system, but with regard to energy, it depends entirely on the body for the exercise of its many and varied functions. The energy it uses is the body's energy. People have no other option but to use the energy of the body. The body is ruled by the psyche here, as it is in all other aspects of the psyche. Without the psyche, the body would be incapable of using its own energy. It would cease to exist as a human.

The psyche uses this energy to play its social roles. Of course, it is society that determines how it will perform these roles. A significant proportion of this performance is executed by habits specified and shaped by society (by the level of civilisation, by the cultural templates related to various classes and specific groupings). For instance, you get up in the morning and wash your face, then if you are a man you might shave and if you are a woman you might put on make-up, then you chose what to wear and have breakfast. What follows depends very much on the society's means of production and level of advancement, and within this, on your class, family, and personal status. You go to a certain school or workplace or take part in a certain retirement activity. All of these actions are undoubtedly defined by society, although it is the body that actually executes them. The psyche can only realise its own specificity and sociality by using the body that carries it.

The simplest way to explain the psychical function of the body is to start from the necessity to discharge-charge with energy, which is a condition of existence for human beings, as it is for any other organism. As it is a definitive feature of the human species, the body can only meet this necessity to discharge-charge within *societal exchange*, in other words in the form of *psychical activity.* All psyches are shaped by social existence. Humans, by definition, always perform the energy exchange necessitated by their own organicism within their social relations. It therefore follows that without social interaction, the human body could not exist. There is no physical motion that is independent of this shaping. An individual who is running is, at the same time, a psyche that is running, because no running style is the same as another. For a myriad of unique reasons, every person who runs does so in their own different way, experiencing different feelings and thoughts as they do so. These specificities are in reality the manifestation of the individual's psyche.

Energy is stored in the body, but it is used by the psyche. Therefore, the feeling of having no energy (lethargy) only originates from the body in exceptional circumstances. Generally speaking, even the most exhausted person knows how to muster enough psychical energy to react appropriately to a societal alert (e.g. very few people, however tired they feel, are immune to the rousing effect of lively music). The main cause of lethargy is psychical. A psyche that has been unable to experience feeltypes in a productive manner accumulates systemic blockages. In these circumstances, the psyche spends most of its energy dealing with the blockages and is unable to get sufficient nourishment from the face-to-face relationships that are essential for healthy energy renewal (discharge-charge). Lethargy, listlessness, boredom, and depression originate more from the inadequacy or unsuitability of face-to-face relationships than a lack of physical energy. Energy might appear to be a most biological element, yet the source and level of individual energy depends on social relationships.[19]

The psyche is an "air" (the same sense of "air" as in someone with too many "airs and graces"); it is a *mode of existence.* Its organic equivalent is the human nervous system. Every nervous system, with all its cells and connections, constitutes a whole. The formation and protection of this whole is dependent on maintaining the formation and integrity of the psyche, which is its social mode of existence. A balanced nervous system requires a proper energy exchange between the organism that hosts it and the surrounding environment. This exchange is a psychical practice and achieved through face-to-face relations. The more these relations gratify the feeltypes, the healthier are the psychical and nervous systems – two faces of the same integral reality. For these systems to function healthily, internal relations must reinforce the balance and strengthen internal congruence. Systemic development occurs through the establishment of new, tighter networks of internal relations, which in turn originate from richer, tighter, and deeper *social relations.*

The fact that the nervous and psychical systems are two faces of the same reality *should not lead to the conclusion that these two systems are one and the same.* The integrity, balance, and development of the nervous system occur through organic elements. The integrity, balance, and development of the psychical system occur through *psychical elements* arising from *social relations.* The template of the psyche cannot be found in the nervous system, but it is true that the integrity, balance, and development of the psyche as a whole rest on the nervous system. It is important to remember that this connection is not between one or other piece of each system, but rather between their entirety. One part of the nervous system does not correspond to another part of the psychical system. The determination is between two whole entities, of which the main determinant is the psychical system. *The active determinant is the psyche*, in other words, it is the input originating from face-to-face relations.

The nervous system can be damaged by an external influence, such as an accident, meaning that the psyche will undoubtedly be damaged as well. But in such a case, it is an exceptional event that has turned the passive determinant into an active one. As a general rule, the psyche is the determinant. The nervous system is as healthy and rich as the health and wealth of the psyche; but the health and wealth of the psyche has a psychical, in other words, societal base. In contrast, the health and wealth of the nervous system is organic and physiological. The health and wealth of both systems are *interconnected*, although *the form of existence* that these two systems take is in no way identical.

The psyche transmits stimuli to the nervous system through face-to-face relations. As a result, the nervous system continually forms new internal links, which carry out the body's energy exchanges through their impulses and enable the body to continue to exist as an organic being. The more positive nourishment social relations provide for the psyche's basic feeltypes, the more the nervous system is strengthened. If input from the psyche is negative and incongruous, it is an organic reality that the nervous system cannot be healthy and balanced. (Positive and congruous input is dependent on face-to-face relations making the individual feel secure without being stifled.) Society determines what form relations take, their intensity, and whether there is interpersonal harmony. The nervous system only *carries the load* of these relations. The more the relations escalate to an intensity that is unbearable for the psychical system, and the more the conflicts are exacerbated, the more the nervous system struggles. Structural defects of the psyche emerge as a malfunctioning nervous system. Just as a dangerous movement can leave our body disabled in some way, a dangerous psychical activity can impair our nervous system. In both circumstances, the disability *materialises* in an *organic form*. Its *origin*, however, is in both cases psychical, i.e. *societal*. This defective pair of systems is reflected on a psychical level by anxiety, neurosis, and psychosis, while on a physiological level it is reflected in various disorders, illnesses, ticks, involuntary motions, etc.

How do we know that the psychical system is weak, imbalanced, and internally incongruous? Above all, it is reflected in the behaviour of the individual. It can also be observed by examining the nervous system, but this can only tell us that the psyche is impaired or ill. While this information may give important clues about the health of the psychical system, it cannot say exactly what the concrete impairment or illness is. The nervous system is an organic reality, and its healthy and balanced formation indicates a healthy, balanced mind.

The nervous system is an inseparable part of the body's system, one of its structural elements, which determines the integrity of the body in various ways and to varying degrees. Anatomy provides a description of this and physiology outlines the way it functions. Awareness of this is crucial to understanding how and in what manner psychical events affect the body. Additionally, the human organism, as a unique organic integral whole, has a series of conditions for existence and growth together with various stages of development, maturing, and decay, which are common to all individuals.

The crucial point when it comes to understanding humankind is the observation that the unique organism, particularly the nervous system, is above all determined by the psychical system. Our movements, which are largely unruly at birth, immediately begin to form a pattern, becoming habitual under the influence of the social interactions around us. These patterns and habits are imprinted into our bodies. This is how we adopt bipedalism, one of the greatest evolutionary developments that make humans human. It is social relations that forge the sounds we can emit into a language pattern, imprinting the language or languages around us into our brain, facial shapes, expressions, and shoulder and hand movements, inscribing them on our nerves, muscle cells, and other cells. In short, as the psychical system forms, in varying degrees it shapes all the subsystems that are the building blocks of the body, especially the nervous system.

To say that everyone has two eyes, two ears, two arms, two legs, one head, and a body is an "obvious" fact that can be *very misleading*. Let us consider the example of two women: one has been virtually turned into a doll by the restrictions and education system of an ancient Chinese palace, such as having her feet bound tightly to prevent their growth; the other, an Amazonian woman, is used to being part-naked, with her skin decorated in multicolour, and has learnt countless uses for a myriad of plants in keeping with her culture. It might be true to say that their bodies are both basically the same, but with respect to humanity they are worlds apart. This difference is created by the psyches of their two separate bodies, which are the products of two very different cultural environments.

It would be wrong to conclude from this that the nervous system and the body it inhabits have a limitless plasticity (can take any form), or that culture (or the psyche that is its individual form) is the only determinant of the body. This is not true. Although the psyche is the active determinant, the body with its nervous system is a passive determinant. It is still a determinant albeit a passive one; it hosts physical structures in various states of vitality, setting constraints and boundaries as it imposes the inevitable energy exchange with its environment on the psyche. Just like the body's other systems, the nervous system has the unique ability to develop, complexify, breakdown, breakup, and repair itself. Development is only possible through the functioning of these abilities. The psyche must conform to the rules of the body onto which it is imprinted. It can only have sovereign power by obeying the objective rules of the body it is subjected to.

7. The Body's Impact

We know that the psyche originates from the individual's societal relations, especially face-to-face ones, and also that the material of the psyche is a feeling-thought combination. We also know that this feeling-thought combination is "conveyed" by the whole body, particularly the nerve cells. Yet, if all this leads us to the conclusion that the body has no effect on the psyche, we would be very mistaken. Such a conclusion could only be the product of a narrow-minded dualistic line of argument, described by Hegel as metaphysics, involving only contradictory notions, either all or nothing, either being or not being.

As previously stated, unless there is some unusual quantitative accumulation or deficiency, the body's impact is passive. This concept of passive impact needs to be thoroughly understood; it is *in no way* tantamount to ineffectiveness, absence of effect or pure passivity. When this is pointed out, it is usually only variables originating from the body that come to mind, such as one body having a structure that for some reason or other has more energy, a faster metabolism than another or has a greater need to satisfy food or sexual stimuli. This state, in turn, also affects the formation of the relevant individual's psyche. That much is true.

But there is a far more important fact that should not be overlooked. The somatic elements that convey the psyche, particularly the nerve cells, are always shaped by the psyche they carry. As the body is shaped by societal relations, it has no elements that do not convey the psyche. Consequently, a purely biological bodily reality unmarked by the stamp of humanity *cannot exist* at any time, under any circumstances. Therefore, physical features passed on through heredity always contain a degree of sociality. In other words, hereditary features are not undefined nerve or muscle cells, they are cells with genetic structures determined by previous societal relations and conditioned to carry certain psychical inclinations, which along with predispositions and conditioning are imprinted onto the individual's genome. So, it should not be surprising that psychological field research shows the hereditary nature of various characteristics. Heritability is not proof that the psyche is physiological. This misconception only arises from seeing the body as a purely organic, unformatted "conveyor" for the psyche. However, in reality, "conveying" is realised through a specific shaping of the body. Bearing in mind that bodily elements cannot exist separate from their forms and that societal relations determine the "shaping" that makes humans human, what is referred to as biological inheritance cannot be separated from societal heredity, especially when talking about bodily elements that convey the psyche.

The main factors determining animal behaviour are the instincts imprinted onto the bodies of each species. It is these instincts that define how they will behave. In some species, this may include some behavioural patterns learnt from their generic surroundings. But whether these behavioural patterns are there from birth or learnt, they apply to all species. The needs of each animal species and how they are met are clear. However, for humans, such pre-determination *does not exist*, because the determining factor in the human species is not the body, but the psyche. That being the case, contrary to animals, we cannot look to generically imprinted somatic traits in order to explain human behaviour. It is true that every psyche can only exist within a body, but ultimately the body is the equipment of the psyche. It is not a determinant but rather it is itself determined. It is important, but *it is not the leader*.

Even the bodily needs considered to be the most fundamental, such as eating and sexual intercourse, can only be accepted as human needs after they have been moulded into a human shape by the psyche. The main determinant imposing this "human shape" on the psyche is the specific culture it feeds off. There is no doubt that the body must continue to live in order for the psyche to exist. But we cannot infer from this that the basic needs of the psyche are the same conditions that maintain the body's life. If that was the case, breathing would be one of the psyche's primary generic needs. As this is not true, we must start out from the psyche itself, not the body, in order to identify the basic needs of the psyche.

The psyche is the individual form of society. It has its own unique reality, just like the body it is hosted in. From birth onwards, this reality is formed by face-to-face relations. It is the relevant culture that determines the nature of these relationships. Are there some characteristics of the psyche's formation that are common to all cultures? Without doubt there are. If the answer to this had been "no", all cultures could not be "human" cultures. Therefore, we need to identify this common "humanity" characteristic; it is, above all, the identical psychical sensibility patterns found in all people, and, as with the body, they developed through evolution. Just as evolution led to our present physical structure around one hundred thousand years ago, the same process equipped our psyche with six basic feeltypes (with affective-intellectual sensibility axes). Although these sensibilities are realised in different forms by different cultures, they are nevertheless the same. This is the physical-psychical singleness of humans as a species. The nature of the being referred to as "human" is the entirety of its societal relations. As these relations vary from one culture to another, so does human nature *vary* between cultures. However, the variability of this active determinant is *not unlimited*. Body-psyche possibilities cannot exceed the constraints of their sensibilities. Despite being *passive*, they are still determinant for humans. The unity of humans as a species originates here.[20]

While of secondary importance, another common characteristic that needs to be emphasised in the human species is the striking similarity in how feelings are manifested. This similarity too is rooted in *evolution*, and therefore, there are similarities with some animal species as well. This subject was first rigorously researched by Charles Darwin in *The Expression of the Emotions in Man and Animals*.[21] As can be understood by the title, this almost 400-page work, embellished with drawings and first published in 1872, is directly dedicated to the subject. In this work, Darwin examines numerous bodily manifestations of mood, such as sadness, anxiety, despair, joy, affection, compassion, adoration, contemplation, nervous disorder, determination, hatred, anger, contempt, guilt, desperation, surprise, fear, terror, and shame, researching how many are found in which animals other than humans. According to Darwin, at the very least, there is a fundamental similarity in the manifestation of feeling among all humans. Darwin explains this similarity in three principles and the first three chapters of the book examine them in turn. According to the first principle, for the manifestation of certain emotions, the movement patterns we display as a result of our physical makeup (e.g. the stance of "being on alert" when sensing danger) become hereditary through repetition over many generations. According to the second principle, again in keeping with our physical makeup, antithetical emotions lead to the development of antithetical movement patterns (e.g. "loosening up" when the danger has passed). These also become hereditary over generations. The third principle arises from our physiological makeup: the stimulation of our nervous system creates a specific series of effects on our skin, muscles, blood circulation, etc. These are reflected in our appearance, stance, and motions, forming a certain manifestation mode. This somatic effect is also hereditary. Consequently, almost all human beings manifest similar moods and neural stimuli with similar patterns. This commonality makes it possible for people to attach meaning to another's appearance. Clearly, without this possibility, societal life, and therefore culture, *could not exist*.[22]

The fact that the body is the obligatory conveyor of the psyche gives other people, and thereby in the final analysis, the social order, the ability to use the psyche to hurt or gratify the body, or to influence the psyche by using the body directly.

We know that the six basic feeltypes, the "senses" of the psyche, are not separate from each other. On the contrary, most of them are active together at any given time. The fact that every feeltype is imprinted into the body means that they are experienced as physical as well as psychical events at the same time. This is obvious in the case of the Pleasure/Pain feeltype. But it is also the case even for the Comprehension/Confusion feeltype, which may seem to be the most cognitive of all. The experience of comprehension at the same time gives people physical pleasure, while confusion creates physical stress. It should not be forgotten that the Like/Dislike, Trust/Fear, Approval/Disapproval, Joy/Sadness feeltypes are always experienced in the form of physical comfort or in the opposite form of tension.

Therefore, feeltypes, which always originate from the face-to-face relations of the psyche's environment, are always accompanied by varying degrees of physical symptoms. By spreading poison or confusion with our words, we can provoke emotions that shake the physical bedrock of others, causing anger, sadness, fear, or shame. This provides other people, and consequently the social order, ample opportunity to bear upon the individual with reward or punishment. And as the feeltypes are unavoidably experienced in the body, it is also where rewards or punishments are felt.

The body's mandatory role as a conveyor makes it possible for others, and consequently society, to guide the psyche by direct interference with the body. One way of doing this is to inflame passive determinants arising from the body by impeding them, and after activating them in this way, using them as a tool of oppression or reward. Of all the physical determinants, the impulse most suited to such use is the sex drive. In this area in particular, societies develop a myriad of taboos that draw individuals into a repressive web. This is why we see the phenomenon of individual sexuality being governed by society everywhere that a state, and therefore repression, exists. The need to eat is another passive determinant well suited to be used as such a weapon. For example, the obligation to work can be tied to the provision of food. The need for motion can also be restricted and turned into a tool of repression. This is the rationale behind imprisonment as a form of scaring or punishing people. Moreover, others, or ultimately the social order, can govern the psyche by more direct means. The most extreme example is the death penalty or killing in general. Killing is the quickest way to destroy another psyche. However, many more types of intervention have been used on the body in order to bring the psyche into line, such as torture, injury, beating, deprivation of food or water, restriction of breathing, bondage, incarceration, and rape. Of course, intervening in the body can also be used to give pleasure. Others, and ultimately the social order, use these pleasures as a vehicle for governing the psyche, for example with feasts or sexual fulfilment. However, this use of the body as a vehicle in societal relations *does not mean* it is a party to or an actor in them. The parties are always psyches. *Actors cannot be made from organisms.*

8. The Role of Sexuality in the Formation of the Soul

The inescapable mediation of the body drew Freud to conclude that sexuality was the most determining human drive. The reasoning is clear: if every feeling-thought is the product of a relationality mediated by the body, sexual relations, the closest type of relationship between two bodies, must be the most determining. Despite appearing convincing at first glance, this notion is fundamentally wrong. It is true that feelings are conveyed through the body. But they are given meaning and concrete content by symbols. Symbols are cultural. The same sexual relationship can be the sign-symbol-indicator of numerous different things depending on the culture or subculture. For example, depending on the culture, it could be a symbol of oppression, repression, exploitation, sin, merit, wealth, slavery, friendship, camaraderie, infatuation, institutional marriage, or victory in battle. The concrete meaning is defined by the culture in which the relationship takes place. The affective power of sexual relations makes them one of the most widespread interpersonal relationship types. That much is true. But the meaning, and therefore affective function, of the relationship is given by the culture.[23]

Undoubtably, in some situations sexual desire can be the most powerful influence on human actions. It is also true that in some societies, sexuality fulfils a particularly important drive-function within the series of standard daily behaviours. Those who make these observations their starting point see sexuality as one of the individual's most powerful biological drives. As known, this view was most strongly advocated under the guise of science by Freud, who also saw this sexual urge as the key to the sociology of the individual. Built on the foundation of sexuality, Eros (Life Instinct) is the source of joy in both individual life and in societal relations. According to Freud, unbridled Eros can lead to societal collapse, so civilisation requires the regulation of Eros. Therefore, in Freud's view, there are two ways in which the existence of civilised society is intricately related to Eros, and therefore sexuality: without Eros there would be no society, and if Eros is unrestricted there would be no civilisation. In short, Eros and sexuality are conditions of existence for civilised society.

A crude supposition lies at the root of this widespread belief, which bases individuality and sociology on biology. According to this, a person can be likened to a bagpipe. Certain hormones of biological origin continually inflate this bag. There is no way to avoid this. If the bag is not periodically emptied by some kind of sexual gratification, it inevitably becomes a source of great anxiety. For this reason, regular sexual gratification is a condition of happiness and health that no individual is exempt from. The sexual urge is biological, existing in the generic human structure. Sexuality is a kind of biological imperative.

This supposition, which underpins numerous views claimed to be scientific, is *primitive* on two levels. On one level it is clearly *patriarchal.* On another level, it is alien to *humankind as a species. The human race* is not a community of stallions whose greatest need is to mate with mares. No-one denies that sexuality, just like hunger, is a drive originating from an individual's biological character. That does not mean that sexuality causes the same behaviours, regardless of historical and cultural conditions. The concrete form that sexuality takes is related to both the social order (culture) and the individual structure (the person's psyche). Except for exceptional situations in which passive determinants have been activated, it is not the biological structure that determines sexuality but rather the culture and the person themselves.

The importance and forms of sexuality vary hugely from one society to another. Sexuality is one of the most effective tools available to societal authorities, and also therefore governments of all types, for keeping individuals under the yoke. Like all drives, the more sexuality is inhibited, the more it is aroused. Therefore, the reason that sexual arousal has become so extraordinarily significant is not due to its qualities, but because it is inhibited by various types of disciplinary regimes. It is unimaginable that sexuality would have its present significance in a society unregulated by a state or institutional authorities. Moreover, in today's capitalist society, sexuality does not just serve the state, but also consumption. As a popular distraction, it counters forms of consciousness that could lead to a change in the prevalent order. As we can see, sexuality can take on different functions depending on the character of the social order. The existence of sexuality is undoubtedly biological. But this biological reality can be utilised for different functions depending on the culture.

A similar variability can be seen in the individual. Grouping by sex or age shows that all individuals do not display identical sexual behaviour according to their cluster. It is normal to find many different sexual behaviours, especially in cultural settings that allow diversity, which vary according to personal history, structuring, experience, and environmental influences.

What shapes the sexual drive is the *object of its attention*. The drive varies greatly according to this object. So much so that the object may not even be another person, for example in the worship of things (fetishism). A shoe or a dress may be the most provocative sexual objects for a particular person. The sexual urge is aroused, weakened, inhibited, or directed by the numerous forms of conditioning that start from the gestation period, particularly the face-to-face relationships that start from birth. As a result of these countless influences, the individual acquires certain tendencies, and the sexual aspect of their personality takes a particular form, but despite the strength of this hold, it is still passive. For example, it may determine whether an individual is heterosexual or homosexual; that is as far as it goes. Within hetero- and homo-sexuality, which individual with which physical, societal, affective, intellectual, and other characteristics will be the object of sexual attention is the result of every individual's unique past and their current lifestyle.

In the randomness of societal relations, the various traits of a particular person, or perhaps the strength of their prominent traits, can be so strong that it makes them the object of passionate desire for another person. As long as the passion continues, this person becomes the unique object of the other person's sexual urge, rendering all other aspects of their individual life relatively insignificant. The person lives, dies, and kills for the object of their individual sexual urge. As the object in every situation is decided by societal factors, it is impossible to explain all this with the power of a biological sexual urge.

According to newspaper reports,[24] US scientists have identified a gene connected to homosexuality in a species of spider. Is it possible to make any conclusions about human behaviour based on this finding? Not in my opinion. Because human sexual orientations are not determined by their genes, but by societal influences that begin in the mother's womb. If that was not the case, a human juvenile growing up alone in a forest would have a sexual orientation of one kind or another. However, sexual inclination per se has not been observed in these living creatures (they are not humans), let alone orientation in one way or another.[25]

When it comes to humans, genetic determination is a passive determinant. Moreover, it is a relatively weak passive determinant. In order for it to become influential, thereby being capable of determining individual behaviour as well, it must be extraordinarily strong. This kind of determinant can be seen for instance in those who are musical geniuses from birth, and as we know, such happenings are extremely rare.

In sexuality, genetic structure can only be influential in specific biological formations. Intersexuality is an example of this. However, even the sexual behaviour of intersexuals is determined by the societal relations of the surrounding environment more than their physiological circumstances.

Humankind is societal. The individual is societal. The individual's sexuality is societal, not biological. Biology is only a condition of *existence* for the sexual drive, as it is for all other drives. Its other characteristics originate from society. The source, form, and intensity of pleasure are all societal. Every love or passion is unique to the person who experiences it; the "explanation" for this is that the individual believes their strongest urges can be satisfied by the object of their sexual urges. Consequently, we sometimes struggle to understand the reasons behind another's desires. The objects of sexual desire are generally determined by the dominant "like" templates of the culture. However, individual preferences have more influence when the society is more flexible. This is where the "unintelligibility" begins. But regardless of whether it is understood by others or not, the basic reality is that, like all biological inclinations, sexual inclination is defined by its object. This object, even if it is a non-human item, is always "societal". Therefore, in no circumstances can sexuality explain society. On the contrary, society explains sexuality as well as all other physical drives.

While all this is correct, there is clearly still a phenomenon in need of explanation: any feelings that gratify the individual, especially enjoyment of life, can often lead to sexual stimulation. This may be the phenomenon that misled Freud and his followers. For example, there is the widely reported observation of mothers feeling sexual arousal while breast-feeding a baby or when pecking its cheek and catching its scent. Other examples are feeling sexual arousal after a notable professional success, making a scientific discovery, breaking a sport's record, or passing a challenging exam. How can these be explained? I think the correct answer is that pleasure experienced in any area is transmitted *through the agency of* the sexual drive. But this should not be *misinterpreted*; it does not mean that all pleasures have a sexual *origin*. If that was the case, the aforementioned feelings experienced by mothers would be classed as incest. The mistake stems from taking the outcome as the reason. Biology frequently reflects our pleasures to us in a sexual form. This in no way means that our pleasures all have a sexual origin. Although it is true that some pleasures derive directly from sexuality, these are determined not by our biology but by societal relations. Just as we cannot help but use our feet for walking, our feelings of pleasure often pass through the body's sexual mechanisms. Just as our feet are not the reason why we walk, so is our sexual drive *not the reason* for our pleasures.

9. The Description of the Soul

Every psyche is a system, and like all systems it must to some degree protect its integrity and congruence in order to survive. At around the age of five, a relatively early stage in a child's growth process, the psyche reaches a threshold. Its determining lines are more or less drawn, an individual ego/personality emerges, and we recognise each other as separate individual personalities. In other words, from the age of around five onwards, the psyche becomes distinct, turning into a separate individual personality. Previously, we briefly looked at the nature of the reality we refer to as the psyche, including its evolutionary roots, formation, structure, and functioning. However, these are each an analysis in their own right. In order to explain different personalities, we need the verbal tools to make their description possible. There are countless adjectives we can use to describe a body, such as young/old, tall/short, dark/fair, etc. So how can we describe a psyche/personality at any of its stages?

After extensive field research and surveys, US psychologists identified five meaningful variables for describing a personality. To facilitate their use, they abbreviated these five characteristics with the word "OCEAN":

"Openness": openness to new things, art, and adventure. *Curiosity.*
"Conscientiousness": sensitive to duty, responsibility, and success. *Responsibility.*
"Extroversion": inclined towards relationships, collective activities. *Outgoingness.*
"Agreeableness": inclined towards sharing, affection. *Kindness.*
"Neuroticism": inclined towards anger, bad temper, and nervous disorders. *Irritability.*

To come up with these variables, surveys were completed in which people were placed in a certain percentile with respect to each variable. *There is no need* for any supporting theoretical framework in order to accept the validity of these concepts for describing and gauging the personality. As implied, these are each just descriptive tools. They do not have an explanatory claim, so they do not need any theoretical justification. Thousands of descriptive words used in the English language were whittled down into these five separate qualifying groups, independent of each other and consistent within themselves, by using surveys to group those with similarities together. These experiments were conducted at different times, by different researchers, in different cultures, and among different age and sex groups, ending up with broadly similar results. Therefore, this list of variables seems to be a reliable tool for describing character.

In their everyday lives, individuals experience thousands of affectivities expressed among the culture and subculture to which they belong. *Which affectivity* they experience is to a large extent determined by the network of face-to-face relations they are involved in, their roll within it, and the structures formed by these roles – in short, the culture around them. *How* the generally accepted affective patterns are experienced is also defined by the culture. What determines how individuals perceive them is each individual's personality.

This personality, which we have seen can be primarily described by five variables ("The Big Five"), is shaped by the everyday affectivities determined by face-to-face relations. These impacts are imprinted onto the body of every human juvenile. As this body is the product of a specific phylogenesis and ontogenesis, it is shaped to some degree even at the time of birth. This personality is continually imprinted, becoming more or less structured over time, although it never becomes entirely rigid. If there are no significant changes to the individual's life, the personality gradually loses its suppleness, becoming increasingly rigid and more stringently determining the way daily affectivities are experienced, thus consolidating its own rigidity. However, when there is a radical change in the nature of face-to-face relations, the personality faces an important test. Success in this test depends on the ability to make adaptations that can prevent it from fragmenting and allow it to maintain its inner congruence. Both fragmentation and fully congruous adaptation are exceptions. The most frequent outcome is a distorted personality that applies a series of defence mechanisms in order to maintain some degree of inner cohesion.

Cultures vary, individuals vary, and every life story is unique. So how can it be that we can describe the personality of most people with just five separate characteristics? To answer this question, we need to grasp three separate factors. Firstly, the five-point description and measuring tool developed by US psychologists is effective for identifying a huge number of different personalities. All five characteristics are applicable to everyone, but this does not mean that ultimately just five personality types are being depicted. In reality, every characteristic is a ladder of many rungs. After the assessment process, every personality is located on a different rung of five different ladders, and with these five separate ladders, it is possible to depict an endless number of different character compositions.

Secondly, as already pointed out, every person arrives in the world with a particular heritage. Since the times of ancient Greece, it has been observed that people can be: sanguine like air, or the opposite, melancholic like earth; choleric like fire, or the opposite, phlegmatic like water. These temperament characteristics are like bodily characteristics, in the sense that the relationship between these characteristics and the psyche is the same as the relationship between the body and the characteristics of tall/short, dark/fair, strong/weak, etc. For the body, these characteristics are not physiological, they are morphological. They are passive not active determinants of the organism, determining its form rather than its nature. The same applies to the psyche and the above temperament characteristics, which are more related to its morphology than its psychology; they are passive not active determinants, determining not what the psyche is, but rather how it is. This does not mean to say that they are insignificant or without influence. When it comes to description, rather than explanation, the temperament traits are clearly functional. Within the five personality traits, it seems that extroversion and neuroticism are most closely associated with the temperament, but the others are not entirely unrelated. For example, being extrovert (like the wind) or changeable (like fire) could be positive temperament traits in an inquisitive personality.

Thirdly, it is not difficult to establish a relationship between feeltypes, which are the sensory axes of all psyches, and the main characteristics describing personalities. However much cultures and people vary, we know that every human has only six sensory axes of the psyche. At the same time, we need to remember that these are experienced through a mixture of feelings, which vary between the two extremes of each axis and also span different axes at the same time. Every axis we use to analyse the psyche, just like the characteristics used to describe personalities, is a ladder of many rungs going from positive to negative. The feelings that the various rungs of each ladder correspond to – Like/Dislike, Trust/Fear, Approval/Disapproval, Joy/Sadness, Pleasure/Pain, Comprehension/Confusion – furnish the personality with characteristics that correspond to the various rungs on the ladders of openness/indifference, conscientiousness/irresponsibility, extroversion/introversion, agreeableness/hostility, calmness/neuroticism. For example, someone who in their concrete life is generally loving, loved, liked, and trusting develops a personality that can be described as more congenial and conscientious. In contrast, someone who is often exposed to painful relationships, rejection, and fear can develop an introverted, indifferent, or neurotic personality.

The essence of the matter is that in order to develop a valid theory about the psyche, it is of *crucial* importance to pay close attention to which factors are *active* and which are *passive*, and to distinguish which concepts should be used for *analysis* and which for *description*.

10. The Specificity of the Individual and the Social

One of the main reasons for mistakes in the theory of humankind is a failure to properly understand the individual-society dialectical identity. As soon as we see the individual and society as two separate realities, we begin to lose hold of the rope. The individual and society are not two separate realities, they are two different faces of the *same reality*. It can be easy to see that there is no society separate from the individual. The main difficulty and source of error is failing to see that there is no such thing as an individual without society. Every individual is necessarily part of a society. Therefore, when we see an individual, we are really *seeing society*.

Failure to understand this can lead even the greatest thinkers down erroneous theoretical pathways. A famous example is Durkheim.[26] He came to prominence for his strong opposition to ideas that tried to restrict humankind to the physiological level. At the end of the nineteenth century, when positivism in the sciences was at its peak, he stubbornly opposed the physiological reductionism that is still widespread today; he argued that humans are above all societal beings rather than a thinking brain. However, his failure to establish the identicality of individual-society leaves him with a dual approach to the individual, on the one hand connecting it with the body and on the other locating sociality in the psyche. This inevitably leads to the viewpoint that humans are beings with an individual physiological side and a social, intellectual, and moral side. In this framework, drives and physiological needs constitute individuality, while intellectual activities and moral concerns constitute sociality. Durkheim argues that the psyche is the individual form of collective consciousness; this characteristic makes it contrary to the body, which forms the basis of the individual. Therefore, human duality is actually the individuality/sociality antinomy.

Looked at in this way, there is no escape from individual-society dualism. Although Durkheim said that sociality always existed in the individual, he could not help but view social existence as separate from the individual. His reasoning is as follows: In nature, higher realities always contain lower realities. For example, a living creature is made up of non-living elements like carbon and nitrogen. But it is a great mistake to let this justify basing the science of upper reality on the science of lower reality. The right solution at every level is to establish the science of upper reality with its own unique concepts. The individual and society are in a relationship of lower reality–upper reality. Just as living matter contains non-living matter, society also contains individuals. Just as living matter cannot exist without non-living matter, society cannot exist without the individual. But it is wrong to try and base the science of society (sociology) on the science of the individual based on this reality. If sociology is to be scientific, it should be exclusive to society.

The reasoning is true from a logical point of view. The mistake comes in seeing the individual–social relationship in the same way as the relationship between a non-living and living entity. Living and non-living entities are by nature generically and qualitatively separate realities. In contrast, the individual is not by nature generically and qualitatively a separate reality from society. *It is identical.*

In actual fact, there is no mechanical, or in the words of Hegel metaphysical, identity. The identity in question is one in which the parties maintain their specificities, or to be more precise, in which they are *only* able to establish their identity *through these very specificities.* Society establishes its own specificity through the *individuality of its individuals.* But this *does not* lead to the conclusion that the individual and society belong to different levels of reality. This is the mistake made by Durkheim and the vast majority of people who hypothesise on humankind.

Contrary to Durkheim, the psyche is not a form of common consciousness embodied in the individual. It is not the case that there is a common consciousness on one hand and an individual psyche on the other. Moreover, there is no such thing as common consciousness. The psyche is the form of existence of an individual's social nature. At the same time, it is the body's form of existence. The human body is not just any biological entity. *Human biologicality is also social.* As previously pointed out, the human body's energy and needs cannot survive without sociality.

As illustrated, the mistake originates from a failure to see the individual-society identicality. There is no doubt that the individual and society are identical, but as stated in the first chapter, this is not a metaphysical but a dialectical identity. This dialectical identity has a metaphysical quality only on its first, most elementary level. In other words, at this level there is a crude identity. When the inertia of this level is surpassed, and the identity of the individual and society can be seen as a process or a formation, it is unavoidable to see this *identity as a dialectical one*, or rather as an *identity containing specificities.* As stated above, the second level of the dialectical identity necessitates exploration of how the parties can function within their own specificities and also be a medium of the other. In practicality, this means a *definition* of the parties. The understanding of identity at its first elementary level does not allow any definition. A definition requires there to be different elements between the defined and the definer. At the second level, the separation of defined and definer is achieved by presenting the specificities of the parties. Therefore, this level requires demonstration of why the individual can only exist collectively and why society can only be realised through the medium of individuality. Hence, the question of this level of identity is tantamount to the theory of the nature of the human species.

The third level of dialectical identity requires a much more detailed analysis, because the subject of research here is how the characteristics that make society and individuals separate and unique can come about through the mediation of the other party (through society for the individual, through the individual for society). For instance, how is it that the ability to feel and think (a condition of existence for the individual) can only materialise within social relations, and by contrast, how is it that the renewal of the forces of production, which are a condition of existence for society, can only be provided through individual energy? The mediations must show how the specificities of individuality transform into the specificities of sociality and vice versa. This means that the theoretical discourse must speak the "language" of social specificities at one end of the mediations and of individual specificities at the other; at the same time, it must retain its own consistency and leave no space for disjunction. Freud's mistake was to start out from the individual and carry on using the same language of the individual when he arrived at society. The same type of mistake is made by anthropologists and structuralists who start out from society, overlooking the specificity of individual structures, and consequently continue to use socio-cultural language ad infinitum. As I have emphasised, the mediations can only be of theoretical value if they can show differential entities transforming into each other. Without this quality, they cannot be true mediations.

The *most fundamental condition* for any valid analysis of the third level of dialectical identity is to ensure that the *historical* attribute of society (and therefore people) is never overlooked. It was no coincidence that Marx referred to social science as "the science of history". The foremost characteristic of the mediations that actualise the transformation of society into the individual and vice versa is that they are historical. They vary according to society, and therefore, according to the historical phase in which they occur. Additionally, it is not just the mediations, but also the specificities of society and the individual that are unique to their historical phase. Overlooking this is one of the main causes of error.

Freud, for example, made this fundamental error by attempting to utilise an analysis of individual psyches that had developed in specific circumstances to explain the individuals of every society. Lucien Sève is also to some degree guilty of the same mistake. Sève pointed out an important issue in individual development by drawing attention to the basic need of all individuals to maximise their skills and use them effectively in social relations, while also observing that failure to meet this need is a major cause of anxiety.[27] The same point was made by Abraham Maslow, who defined the greatest individual need as the need for self-realisation.[28] These determinations, reached by Sève from the starting point of society and

Maslow from the starting point of the individual, are correct. But they are valid in certain conditions: the conditions of advanced capitalist society, or in other words at advanced stages of individualisation-socialisation. Their mistake was to defend its validity in every type of society and individual, thereby overlooking their historical specificities.

The first chapter on cognitive tools points out that the outcome of any determination generally depends, not on the determinant, but the determined. More precisely, the existence of an outcome depends on the determinant, and the quality of it depends on the determined. According to historical materialism, the main determinant is the mode of production, which defines the whole society and therefore its individuals too. However, according to this rule, the same general determinant can only show its impact in the framework of specificities of the fields it has determined. Therefore, the outcome of the determination must necessarily manifest itself with the "language" of the specified field. Consequently, the sociological determination will be wrapped in a psychological form at the individual level. In analyses, it is *wrong* to use concepts specific to the psyche for the body, concepts specific to the body for the psyche, concepts specific to society for the individual, and concepts specific to the individual for society. Every concept must be expressed in the language of the field it belongs to, regardless of the determinant.

This question of method is one of the *crucial* questions of human theory. The two main sources of the greatest errors in human theory are: firstly, trying to explain individual behaviour using historical materialism's laws of social change without applying the language of its own specific fields; secondly, trying to analyse social development with the language of psychology rather than the language of its own unique field. In their own fields, neither the rules of societal development contained in the materialist history approach nor psychology's rules of individual development are wrong. But when one is transferred, as is, to the other's field, their *verity* is lost, *leading to major errors*. Theoretically, the only valid solution is to grasp the *mediations* that transform the truths of one field into the truths of the other.

However, it is equally misguided to claim that, until these mediations are found, no valid scientific analyses can be made at the social or individual level. The opposite is true. Both levels must be researched with their own specific conceptualisations, on condition that their identicality is always borne in mind. This is no place for cutting corners. Unless society on its own and the individual on its own are thoroughly studied within their own specificities, focusing exclusively on the mediations between the levels would mean ignoring the fundamental truth that identity can only be grasped through specificity. Neither Marx nor Freud lived in vain.

From the point of view of individual existence and development, or put more simply, of being an individual, social relations above all mean face-to-face relations. Every individual might form relationships, whether affective or cognitive, with at most 100 to 150 other individuals. These interactions shape the individual and also make them who they are in the myriad of societal roles. No individual can be understood without defining and analysing these. Therefore, when moving from the social to the individual, it is crucial not to jump from general society to the unique individual, and it is essential not to overlook the fact that face-to-face relations are *essential mediations* for understanding and explaining the individual.

When it comes to social existence, it is the relations that turn individual energy into social force that are important, rather than the face-to-face relations that shape the psyche. These are chiefly the relations that enable the reproduction of society's material existence, but also the structures that maintain the integrity of society. While it is true that the existence and development of an individual means the existence and development of an individual identity, from the point of view of society, it is of little importance which concrete identities undertake the roles that enable the social order to exist and develop. The reason I have said "of little" importance rather than "of no" importance is that at crucial mediation points in individual-social transformation, and depending on historical conditions, there is sometimes the potential for the social structure to acquire greater importance for the individual, and equally for the individual identity to become more significant for society.

In individual-social relations, the main determinant is society, regardless of time and conditions. This is why humans are generically social creatures. We cannot explain society through the individual, yet we can *only* explain the individual through society. Society's need for the mediation of individual specificities also changes according to the conditions of the society. In stagnant periods, society's need for this is at its lowest level. Yet in certain historical conditions, individual specificities gain great significance for the functioning and, if necessary, the transformation of society. When social development stagnates, the only way for society to overcome the problem is the emergence of individual attitudes that can effect change. Society can *only be changed by the individual*, not by society itself.

As we can witness today, when social change has become a life and death matter for society's existence, the only solution is to trigger individual consciousnesses that can bring about transformation. For an increasing number of people, factors like environmental destruction and climate change have made transcending capitalism *an individual need*. Whenever this need becomes sufficiently widespread and concentrated, society will have provided the individual mediation necessary for its own salvation. In such an environment, individual actions will turn into the most powerful social tools. The individual determined by society will have become society's chief determinant. But we should not be misled by this. What decides whether the individual is important, and how their impact will operate if they are important, is *always society*.

Section B: Literature

1. Feeltypes and Piaget's Theory of Soul Formation

The Swiss scientist Jean Piaget (1896–1980) appears to be by far the most cited writer in the *Handbook of Child Psychology*, which entered its fifth edition in 1998. First published in 1931, and subsequently reprinted every few years, it is not only the most important publication in the field of child psychology, but also one of the most impressive undertakings in psychology as a whole.[29] Therefore, being cited in this work is a fair measure of a writer's status, and with an output of over 50 books and 520 articles, Jean Piaget is indeed one of the leading names in cognitive development theory.[30] Here, I will neither summarise nor assess Piaget's substantial contribution. My aim is simply to research the expediency of Piaget within the context of the feeltype theory advocated in my work.

As already stated, the basic assertion of the feeltype theory is that the human psyche is established along six main axes: Like/Dislike, Trust/Fear, Approval/Disapproval, Joy/Sadness, Pleasure/Pain, and Comprehension/Confusion. Inferences to every one of these axes can be drawn from the thought system created by Piaget, but even more important than this was his examination of the developmental process of child intellect. Hence, he referred to his theoretical framework as "genetic epistemology". Of all the areas of a child's psychical formation, Piaget focussed primarily on cognitive development. In my opinion, his success here has been unsurpassed to this day. As can be seen, Piaget's theory mostly relates to the last of the six feeltypes listed, and it is important to bear this proviso in mind when examining his contribution.

According to Piaget, intelligence is a product of dual processes that continually nurture and condition each other: assimilation and accommodation. Assimilation is the internalisation originating from the endless flow of activity experienced by a child. In other words, it is the impression left on the child by their activities. But this impression is not stamped onto a blank canvas as there will always be a ready-made structure formed by previous experiences. With every impression stamped onto it, the structure changes accordingly. This process is called "accommodation". Therefore, learning is never passive – it always occurs within an activity. Learning only comes about *through changes* in the object, not through observation. This change comes about either through the physical properties of the object (the child bites or breaks the toy, sees inside it, etc.) or through the child changing its position in order to compare it as part of a set of toys (the child makes the toy part of a set of toys).[31] In both cases, there is a changing action. What is learnt is not the object being observed, but this *action* of changing. This is how assimilation occurs and leads to new accommodation. The next assimilation is then subjected to the action of this changed structure, which has readapted as a result of learning. In short, learning occurs not through the generalisation of perceptions as was generally thought, but through the generalisation of experiences.[32]

Piaget examined the childhood developmental process through countless observations and experiments, eventually arriving at a development stage model common to all humans. This process was divided into three significant stages: (1) sensory-motor scheme, (2) figurative scheme, and (3) operational scheme. In the first period, which lasts until the age of around eighteen to twenty-four months, the child acquires a framework of mechanisms that can be described as practical intelligence. The second period continues until about seven or eight years of age, and during this time, the child uses the same mechanisms with figurative tools. The child begins to use the common language of its environment to reconstruct their practical intelligence with symbolic tools; in other words, they give meaning to it. In the third period, which continues until the age of about fourteen or fifteen, the child learns to move from symbols to concepts, again in a graduated developmental process, thus acquiring the determining feature of intelligence: the skill to firstly mentally form an aim, use concepts to choose the most appropriate tools for the aim, and then take action.

Not content with this segmentation, Piaget broke each period down into further stages in order to examine them further. For example, the sensorimotor modulation period is divided into six stages. A child comes into the world equipped with a series of reflexes, such as knowing how to suck and grasp with its hands. However, in a very short time, these natural reflexes are refined by conditioned reflexes. For example, the child learns how to use its fingers to move the nipple so it is fully over its mouth. After a few weeks, the child enters the second stage. At this stage, through the accumulation of conditioned reflexes, the child obtains behavioural patterns that might be described as a set of habits. In the first months of life, the child has no concept of objects. It exists in a kind of magma of which it is a part. The third stage of development is reached when the child is from four to four and a half months old. At this stage, the child begins to combine and coordinate its abilities to see and to hold things. The fourth stage, which lasts from five to six months until ten months of age, is extremely important. The child makes random attempts to achieve certain goals by using its ability of sight and grabbing whatever is around. For example, it tries different motions to get a sound from chimes hanging over its cot or in order to find an object it has seen and lost (a toy, teat, etc). The child is now trying to achieve a goal that is related to objects and has therefore crossed the threshold to practical intelligence. At the fifth stage, which begins at eleven to twelve months, the child acquires new practical modalities through trial and error. For example, pulling the rug closer to gain access to an object on it. At the last stage, which lasts from about eighteen months to twenty-four months of age, it creates new modalities from the behavioural patterns it has already acquired in order to achieve a certain aim. For example, after attempting to reach the contents of a half-open matchbox for some time, it stops and then opens the box by sticking a finger into the gap. This is the first step towards insight and abstract intelligence, and it brings the sensorimotor modulation period to a close.[33] During this period, the child has acquired the fundamental dimensions of the relationships of objects to the world: object permanence (the fact that the object stays the same even if it is moved, or has been out of sight for some time), space as the context of the object, time consecutive to the movements of the object, and the perception of causality leading to the movement of the object. However, these are still raw. For example, the child initially understands causality with a magical-phenomenological quality. This is because it takes its own action (the magical aspect) and the proximity of one object to another (phenomenological aspect) as the source of causality. But the important point is that in the sensorimotor modulation stage, the child has learnt the essential patterns of relations with the world of objects, albeit in an operant way.[34]

Like the first period, the second is also broken down into increments. It begins around the age of eighteen months, when the child first begins to pick up language, and continues until the age of around seven or eight years. The child experiments with the symbolic data they have just begun to grasp. This development can be likened to the progress made in the first three or four stages of the sensorimotor scheme. At the first stage of the figurative scheme, the child is still swimming in a magma of symbols but begins to distinguish symbolic behaviours through deferred imitation. Then, the child establishes certain links between symbols (or more precisely, the images of them in the child's head) through symbolic games. Later, the child attempts to create new figurative schemes itself through experimenting with scribbling and drawing. The child gradually perceives the symbols as image mechanisms. At the same time, speech is developing and the child learns to verbalise the symbolisations.[35] At this last stage, there is a transition from symbols to indicators, or in other words, artificial representations independent of what can actually be seen. But as comprehension of these indicators depends on others attaching the same meaning to them, learning how to use language properly makes the child party to a particular cultural environment. Once this stage has been completed, the child no longer sees themselves as the centre of their entire environment, understanding that they are an object in a world of objects that includes others.[36] Piaget suggested that children's egocentric talk ends at this stage. Yet, according to Vygotsky (who will be discussed later), the egocentric talk develops further and turns into an internal conversation. This internal language enables the child to establish their inner world in a gradually more conscious way.[37]

The operational scheme, which lasts from ages seven or eight until age fourteen, is split into two sub-periods. Up until pre-adolescence at the age of about eleven or twelve, children begin to experience the fourth-fifth-sixth stages of the sensorimotor scheme on a *figurative level* (Piaget labels this sub-period as "concrete-operational"). In other words, they initially try various modalities to solve a problem, then begin to choose between them, and at the last stage they start to seek the appropriate figurative modalities for problem solving. This brings children to the threshold of conceptual thought. Concrete perceptions that are acquired practically in relation to the world of objects during the sensorimotor scheme period, such as identity, combinability, reversibility, additivity, and sequentiality, turn into *general intellectual abilities.* Children learn conceptual thinking at the age of fourteen or fifteen, a universal development stage of humanity. In contrast to their level of perception at the ages of seven or eight, concepts then acquire an increasingly *abstract quality,* separated from reality. *An adolescent first defines an aim in their mind, then uses the concepts in their mind to decide which tools are necessary before consciously taking an action.* Thus, they have reached the *universal stage of minimum intellect* that defines the human race. Some societies never get beyond this point. Those who do so establish the subsequent intellectual levels in non-universal ways that are specific to their culture.

How can this theory, which Piaget developed for intellectual development, be adapted for the psyche as a whole? Crucially, although the mind is a feeling-thought system in which feeling and thought are always nestled together, feeling is the main determinant feature. Therefore, a theory based on thought can only be used to explain the entirety of psychical development if the specificities of the feeling feature are kept in mind. In my view, *two main points* on which Piaget's theory are grounded are valid for the development of the entire psyche: firstly, his views about *learning,* and secondly his *graduated development model* of three stages broken into substages, starting from birth and continuing until the age of fourteen or fifteen.

For Piaget, learning is a dual process based on the union of assimilation and accommodation. Children learn not from their senses but *from actions* in which they use their senses. Action is always an undertaking that affects the object. With every concrete undertaking, the object is *changed* in some way or the other. At the same time, the changing action becomes the source of acquiring a new skill. Skills learnt in this way are added to previously formed behaviour habits, and in the process the whole system changes, completely integrating the new skill. As a result of this accommodation, every modifying action forms a new system, which is a modified form of the former one. The next modifying action will be experienced by this new configuration, and with every newly acquired skill the system is updated once again. This is the learning process.

The process does not just apply to the development of intellect, but to the development of the entire psyche. As well as using their sensory capabilities, children also use their affective capabilities (feeltypes). They exhibit different behaviours in their exchanges with those around them.

Depending on their stage of development, children may cry, shout, stare, smile, grasp, reject, etc. These are the resources of the child's affective learning, which occurs through *symbolisation*. The acquired inputs enter an affective system that is constantly being updated, invoking another change in accordance with the new input. This brings about an affective accommodation. The next affective behaviour is managed by this updated system. New experiences lead to new assimilations, which in turn lead to new accommodations. This is how the child's psyche is formed. Just like the acquisition of practical habits that inform the intellect, this affective intake is initially entirely unconscious.

Piaget's graduated development model with three stages broken down into sub-stages can be applied to the entire psyche. The first stage, from birth until about eighteen to twenty-four months, is the *sensorimotor* scheme, which itself passes through six different stages. It can thus be seen that for Piaget a scheme is not a fixed form, it is a process of generalisation through experience. In his own words: "The history and development of a schema therefore consist primarily in its generalization, through application to increasingly varied circumstances."[38] Clearly, a scheme is actually a behavioural pattern that has become habitual. Referring to it as a "skill" may make the concept easier to understand. When looking at the affective aspect of development, we might refer to this period of a child's development as the *feeling-behaviour* scheme. As with the sensorimotor scheme, children come into the world with a natural reflex capability, albeit small. The reflexes of seeking fullness and crying when hungry are affective capabilities that correspond with the abilities to suck and grasp. Within a very short time, children acquire conditioned reflexes as a result of the behaviours of those around, obtaining a certain habitual framework up until the age of four or five months. Sucking, crying, and sleeping form a certain scheme. As with the sensorimotor scheme, the period from five to ten months is crucial. Just as children at this stage try various motions in conjunction with their visual ability, they also experiment with different *affective tools* to see how much gratification they can achieve and which obstacles they can overcome. Thus, the various achievements assimilated from behaviours like smiling, crying, shouting, looking, gesticulating, and crawling are accommodated as features of an affective system. The child has now gained a particular behaviour set. Then, until the age of eighteen to twenty-four months, they begin to pick out the appropriate behaviours from this set through trial and error, eventually learning the behaviour best suited to achieving the desired result. (As will be emphasised later, first by Vygotsky and then by Mead, this achievement also calls on a process of *symbolisation. No comprehension or learning*, even unconscious, can occur without the process of symbolisation.)

From around eighteen to twenty-four months to seven years of age, children go through the *figurative* period in stages, gradually learning to speak, firstly with single words and then moving onto short phrases. Initially, they are hardly able to discern their feelings and experience them all together in a jumbled mass. During this process, children imitate and repeat the behaviour of others, playing symbolic games that display certain feelings and making scribbles and drawings that are practical expressions of their emotions. Over time, they make various associations to represent different feelings, which in the final stage, they begin to express through language. In this way, by the time children reach the age of seven or eight, the different feelings they experience, such as like, trust, approval, joy, pleasure, or their opposites, such as dislike, fear, shame, sadness, and pain, are differentiated in their own minds, although they are unaware of it. Children gradually shed their egocentrism and by the end of the same period, they experience their own emotions through an exchange with the feelings of others. Moreover, they have learnt to verbalise this affective exchange albeit in a somewhat fledgling fashion.

The *operational* stage is made up of two periods. From age seven or eight to about ten or twelve, children try to understand their own feelings through internal speech. The ego develops through this process, but the child's affective world is largely subconscious. By the time they get to the age of fourteen or fifteen, they are beginning to define their emotions with concepts, thus embarking on the stage of personality development.

At the end of the third stage when children reach puberty, although they may not be able to expressly define them, they have acquired the ability to *use the words of their own culture to affectively-intellectually recognise the concrete feelings and compound feeling-combinations* that are located along the axes of the six fundamental feeltypes shared by all humanity: Like/Dislike, Trust/Fear, Approval/Disapproval, Joy/Sadness, Pleasure/Pain, and Comprehension/Confusion.

As can be seen, in terms of the qualities of the learning process (assimilation- accommodation) and the child's developmental stages and substages, Piaget's cognitive development theory can be applied to all psychical development. But however much feeling and thought in humans are always to some degree combined, it is important to remember that affective development (the psyche) occurs through *separate pathways* to rational development (logic), ultimately forming a *different framework*. The psyche and logical thought *are separate structures*. The building blocks of the logical system are the object's permanence (identity), perception of space and time and causality relations. In contrast, the building blocks of the affective system are the affectivity modalities universalised by the six sensibility axes of the human psyche, and they are always realised through the words of a concrete culture. The purpose of the logical structure is to make changes to the world of objects. The success of the changes depends on the rationality of the concepts. The purpose of the affective framework is to influence inter-individual relations and to bring about changes to them. The success of the changes is connected not to their logicality but to their affective expedience. Although the logical structure may not need consciousness to form the habits on which concepts are based, the concepts themselves are entirely located in the *sphere of consciousness*. In summary, while feelings and thoughts in the human psyche are always to some degree intermixed, the aims, tools, structure, and operational rules of the affective system are separate (different) from those of the rational system.

In the affective world, children are more concerned with people than objects. Their aim is to meet the needs of their affective structure. For example, during the figurative scheme stage, the cognitive and affective functions of the imitative behaviours displayed by children are completely different. Repeating the behaviour of others enables children to form figurative equivalents to the skills they acquired during the sensorimotor stage, thereby enhancing their cognitive skills. However, the affective function of imitation is altogether different. It enables children to identify with things they like, draw attention to behaviours they dislike, or gratify other subconscious needs.

The sole aim of the affective world is emotional gratification. When children emerge into the world, they establish relationships within the dualism of seeking pleasure and avoiding pain. Every experience restructures their affective world again. But this restructuring is very different to the scheme Piaget used for cognitive development. The child's main concern here is not understanding the world, but to perpetuate and maximise their own affective gratification. Like every organism, a child must continually expend and renew inner energy. Both these processes are a source of pleasure and are always realised through relationships with others, relationships that are specified by society. By constantly seeking gratification in this environment, the child induces the shaping of their own array of feelings. Whenever this gratification is confronted by difficulties, obstacles, or pain, children use the possibilities available to them, or develop new ones, in order to achieve gratification and protect themselves. These efforts are at the same time supported by the *affective symbolisation* mechanism the child is just beginning to develop. Eventually, the child's psyche is formed through this process, within which logic sometimes has a positive effect on emotional gratification. The more that logic is instrumental to gratification, the greater is the consciousness acquired by the psychical framework. On the other hand, the unconscious area is expanded when the opposite is true. Evidently, this affective processing model *bears no relation* to the cognitive processing model developed by Piaget.

Ultimately, Piaget was wrong to claim that the cognitive development beginning with birth and ending in adolescence was universal and that only the following stages were shaped by culture. It is not just the means-modes, objects-aims of emotional gratification that are societal, but also those of the cognitive interaction related to objects. They are all culture specific. As already stated, it is wrong to jump to presumptive universal conclusions based on a few facts, such as the fact that humans are organisms. Universal reflexes like hunger or sexual attraction always materialise in a form shaped by a specific culture. Similarly, it cannot be assumed that there is a valid common cognitive development process for all societies, as if the maxim, "Great minds think alike", was literally true. However, it can be assumed that the logic/intelligence of all humans, just like their bodies, is made up of common elements, while its concrete operation will nevertheless be *different* from one culture to another. A child can only undergo cognitive development through the lens of specific templates outlined by the society it lives in, as is true for emotional gratification.

2. Lev Vygotsky's Cultural-Historical Psychology Approach

Vygotsky was born in the same year as Piaget (1896), but he died of tuberculosis in 1934 at the age of just thirty-eight. Consequently, in comparison to Piaget, who died at the age of eighty-four in 1980, his work took place in a much shorter timeframe. In fact, Lev Vygotsky packed the fruits of his work in psychology into the last ten years of his life. Due to his health and the conditions in the Soviet Union at that time, much of his works remained unpublished. Those that were published were not reprinted as his views were contrary to party thinking and incurred the wrath of the dominant Stalinism of the day. In 1965, after the Soviet ice had begun to melt, his first great work *The Psychology of Art* was finally published in the Soviet Union. It was not until the 1980s that all his works finally saw the light of day.

Like Piaget, Vygotsky believed that children learnt societal life through the process of activities and that this learning process was continually restructuring the child. But unlike Piaget, he suggested that adults played an essential role in this activity process, even arguing that children can only experience the sensorimotor scheme through the mediation of others and not of their own accord. As adults are always in a child's "zone of proximal development", they pave the way for their potential development at every stage of their growth. Contrary to Piaget, the individuation process is not physiological and therefore not universal; it is specific to the particular society, or in other words, a cultural process. Children only understand the meaning of their own actions through the light of their relationships with adults. When engaged in various activities, children use the tools of meaning presented by adults to give meaning to their relationships with objects and other people. Whether at school or at home, children can only acquire a world of meaning and comprehend their environment with the guidance of the adults around them.

Vygotsky gives the following example: a baby who wants to reach an object in the distance makes a "grabbing" motion with the hand. Seeing this, a nearby person – such as their mother – picks up the object and gives it to the child. As this exchange is repeated, the child begins to use this particular form of reaching out with their hand as a sign of wanting the object they're looking at. According to Vygotsky, all psychical functions, from the simplest to the most complex, develop in this way. The space in which they all develop at an individual level is the brain and the nervous system; but the origin of their development is societal. This even applies to human volition; the child learns how to impose their will from the conflict of contrasting desires in social exchanges. In summary, all psychical functions initially emerge within the social exchanges that the baby-child experiences. The child always develops them through *internalisation*.[39]

By contrast, the starting point of the traditional approach is the individual's biological structure. The possibilities inherent in this structure develop through societal exchanges; the child is "socialised". This is the view of Piaget, who believed that children develop through practice and experience. However, Vygotsky argued that children develop through *individuation* within social exchanges, rather than socialisation. The process begins with society; the means of understanding (cultural tools) stemming from social exchanges establish the individual intellect by shaping their biological potential.

During this learning process, the child's biological psychical structure undergoes a radical transformation. This is made possible by language. Except for the "special status" of proper nouns, every other noun refers not to a single object but to a cluster of objects. Therefore, from the very beginning, words catalyse a process of *generalisation. This is qualitatively different* to the primitive mind's comprehension through sensation and perception; it is a much more advanced form of comprehension. Consequently, it is a thought practice.[40]

According to Vygotsky, in both phylogenesis and ontogenesis, thought and speech are two separate processes. Language is a combination of both processes. When a child learns to speak, their elementary thought structure is built on a framework of meaning presented by society. However, contrary to the thoughts of Piaget, development cannot start from an introverted ("autistic") stage based on the child's subjective fantasies – the stage of language that he called "egocentric". For Vygotsky, the development curve is the complete opposite. In both children and adults, the primary function of language is communication, it is societal exchange. Therefore, the language first acquired by the child is societal not personal. Initially, this language is multifunctional and wholesale. Functional differentiation only occurs through a process. At a specific stage of development, children separate this social language into two languages – a communication language and an egocentric language.[41] Egocentric language undergoes a particular process to turn into an inner speech that shapes the child's thoughts for them.

The fundamental misconception of both Freud and Piaget was to set the child's intellect upon a standalone primitive, biological foundation. According to this approach, societal factors are "alien" forces that subjugate a child's biological structure.[42] The child's sociality is a feature that is *added afterwards*. Whereas in reality, a child is societal from the start and can only establish their own psychical framework by *harvesting* from sociality.[43] At around the age of two, children begin to combine meaning and sound and to perceive words as carriers of meaning. Prior to this stage, children achieve some elementary intellectual actions such as recognition or recall, but use sounds only as affective tools. The transformation of sounds into an intellectual tool, the perception that objects have a "name", is the first step of generalisation and symbolic thought, which makes it *extremely important*. At this first stage, "grouping" functions as a sign of "relatedness", just as with surnames or family names. But once this connection attains a rational quality, conceptual language is born.

When a child acquires language, their intellectual development becomes connected to the language and cultural experiences available from society. For example, conceptual language can only be established with the specific generalisation norms presented by social exchanges. Being able to formulate concepts depends on the ability to use words as a "tool".[44] It is not until adolescence that children develop the ability to perceive conceptual language. Adolescence is an essential stage of biological development that is necessary in order to grasp concepts. But conceptual language itself undoubtedly has social rather than biological roots.

Conceptual language could not be suddenly acquired in a fully matured state. Mastery of conceptual language develops alongside the development of the individual's level of consciousness, powers of abstraction, and observational skills. The most important tools in conceptual thought are properly managed *generalisation-abstraction* processes. It is only if these are successful that the processes of inter-conceptual equivalence, of comparison, of making judgements, and of forming conclusions can function healthily.[45] The core relationship between different concepts is the relationship of *generality*. The attributes of conceptual language also determine its logical variables. New words are created, and existing words obtain new meanings. The abstraction level of elementary generalisations advances until the most *abstract generalisations* are attained. Language that changes in this way re-establishes its meaning structure and is at the same time internalised by individuals with this more advanced structure. The process reshapes their psychological frameworks.

Like Mead (as we will see later[46]), Vygotsky explains the formation of personality through the internalisation of various cultural tools. Again like Mead, he emphasised that the "meaning structures" ("meaning patterns" established with symbolisation and therefore with societal tools) acquired from social relations are a factor in the formation of these roles. However, he differed from Mead on two important points. While Mead treated these patterns as fixed data, Vygotsky pointed to their *historicity*, reiterating that cultural forms and their content vary according to the historical era. Consequently, his approach was called "socio-historical". Vygotsky's second great contribution is that, contrary to Mead who tended to focus on the social aspect of the individual–society relationship, he tried to explain the individual *personality development process*, thereby developing his own unique "psychological" theory.[47]

At the heart of Vygotsky's psychological theory is the view that human consciousness is related to the use of "psychological tools" or "signs". Indeed, humans establish all their relationships, whether with the physical environment or other people, using the possibilities presented to them by symbolic tools. This is the only way that they can behave "meaningfully" towards themselves or others. Humans can only interact with their environment through the *mediation* of the tools of their culture. The most important symbolic tools developed by humans up until the present day can be summarised as follows: language, counting systems, recollection techniques, algebraic symbol systems, artistic creations, writing, various graphs, charts, maps, drawings, and a range of commonly accepted artificial signs.[48] The crucial point of the theory is this: using the tools that are readily available in society adapts the developing psychical structures *to the culture itself*. Every cultural tool, every sign system, leads to *qualitative changes* in the person using it. In this respect, *learning* amounts to people acquiring the skill to use certain *sign systems*. The more an individual acquires the skills, the more *power they have to influence* their physical and social environments. The environment that has been transformed by this influence is internalised by the individual in its new form.[49] This mutual interaction continues until the individual dies. History shapes the psyche through the medium of culture, while the psyche makes history by using the medium of culture.

Clearly, Lev Vygotsky's approach to the individual psyche was entirely in keeping with that of Marx. Vygotsky himself mentions this concordance, attributing the basis of his view of the psyche to Marx's theory: "But the essence of man is no abstraction inherent in each single individual. In its reality, it is the ensemble of the social relations."[50] According to this, the nature of the human intellectual framework is best understood as the internalisation of all social relations by individuals, who then transform them into individual functions and forms. In other words, at all stages of history, individual psyches form through the internalisation of cultural tools developed at a social level.[51] This formation process is at the same time equivalent to psyches (individuals) "making their own history".

The final question we need to ask is this: how can the feeltype theory and Vygotsky's socio-historical approach be latched together? Vygotsky's theory, just like that of Piaget, was a theory of acquiring *cognitive* ability. Vygotsky went further than Piaget by freeing the theory from a biological-universal base and instead connecting it to historical development, but both restrict themselves to investigating the cognitive aspects of the psychical structure. However, as we know, it is the affective aspect of the psyche that is dominant regardless of how much it consists of a feeling-thought mixture. Just as with Piaget's theory, the feeltype theory makes it possible to apply Vygotsky's cognitive approach to the affective formation of the psyche. The task that lies ahead is to research how each feeltype is shaped and expressed in the relevant culture in every historical stage and environment and within the meaning patterns of common language. Therefore, it is not just the formation of the Comprehension feeltype in a particular historical stage and environment that needs to be developed, but also that of the other feeltypes of Like, Trust, Approval, Joy, and Pleasure. Thus, the "individuation" mechanisms that Vygotsky researched solely for the psyche's cognitive aspects can be determined for all axes of the psyche.

It is particularly important to note that the feeltype theory can help to overcome a common difficulty encountered with all theories in the Marxist framework, including that of Vygotsky. All these theories rest on a specific understanding of *historical development*. But this general understanding is not sufficient for grasping a particular historical stage and environment in its fullness. It is also necessary to construct a detailed, comprehensive theory for each particular stage. The most outstanding example of this in Marxist history is the analysis of the mode of capitalist production developed by Marx, who established an all-encompassing conceptual framework and also concretised it in a specific way. This approach must be broadened. Specific theories and analyses must be developed, aiming at the historical concretisation of different planes of social life, such as politics, ideology, art, etc., while staying within this comprehensive framework of understanding. One of these is the culture-psyche plane. To date, important contributions have been made to the theories of various other planes, and I believe the feeltype theory will be a valuable contribution to the culture-psyche plane of individual/society research. As we study the culture-psyche plane in various historical stages and environments, following the tracks of the concrete configurations of the same feeltypes in each one, it gives us the opportunity to illuminate the culture-psyche reality of the relevant stage and environment in such a way that it can be compared with that of others. This could be the contribution of the feeltype theory to Vygotsky's approach.

3. Foremost Personality Theories and Marxism

According to Jerry M. Burger, who wrote a very informative handbook on the subject of personality:[52] "Personality can be defined as *consistent behaviour patterns and intrapersonal processes originating within the individual.*"[53] These patterns are continuous forms of behaviour displayed by the individual more or less consistently throughout their life. Intrapersonal processes are the motivational, affective, and cognitive processes within each individual.

So, what are the sources of "consistent behaviour patterns and intrapersonal processes"? According to Burger, to date six basic approaches to this subject have been developed. All personality theories can be located within one of these approaches. "These are the psychoanalytic approach, the trait approach, the biological approach, the humanistic approach, the behavioural/social learning approach, and the cognitive approach."[54] Clearly, the historical materialist (Marxist) approach has been left off the list. The cultural psychology approach is also missing, although Burger stressed the importance of cultural factors, saying: "… Psychologists have come to see that people and their personalities exist within a cultural context. Perhaps the most important distinction cross-cultural researchers make is between individualistic cultures and collectivist cultures."[55] For example, this distinction leads to significant differences between cultures when it comes to definitions of self-worth or success. Nevertheless, Burger overlooks the original approaches to understanding personality offered by both historical materialism and cultural psychology.

Despite these shortcomings, Burger's overview of the literature was quite comprehensive with respect to personality theory. Burger's groupings are useful for the purposes pursued here and there follows a short précis of each of the six approaches, listed under separate subheadings. Firstly, the main arguments on personality will be advocated for each approach. They will then be critically assessed with two aims: if the argument is contrary to the Marxist approach, the criticism will help to clarify this aspect of the historical materialist approach; if, on the other hand, the argument is in keeping with the Marxist approach, the historical materialist approach to personality will be enriched by its contribution.

a) The Psychoanalytic Approach and Marxism

The main arguments of the psychoanalytical approach advocated by Freud can be summarised as follows:

First argument: the human psyche is a system or structure primarily consisting of three parts – id, ego, and superego. People are largely unaware of most of the feeling-thoughts that make up the structure. They remain subconscious and only a few can be easily summoned into consciousness. Very few are clearly recognised. The main factors governing human behaviour are these unconscious elements, which originate from two sources; most importantly, they are drives arising from the individual's own body. Everyone has two basic instincts – life instinct and death instinct.[56] The life instinct is driven by "libido". This surging life-energy is basically sexual. It recognises neither time nor place, seeking instant satisfaction. It is governed by the "pleasure principle". The second basic instinct is death instinct. The individual inwardly longs for the calmness of death in order to escape from the stress of life and this is outwardly displayed by aggression towards others. From birth onwards, these two instincts are confronted with the prohibitions and feelings of shame constructed by the person's environment. The individual eventually finds ways to satisfy their sexuality and aggression in forms that will be accepted by society. This task falls to the conscious part of the mind. Consciousness tries to achieve a balance between instincts and the necessities of societal life by following the "reality principle".

Second argument: starting from birth, every child undergoes a psychical development process that corresponds with the region of the body that has become sexually sensitised. These steps, which continue until the ages of about five or six, are oral, anal, and genital stages, respectively. In this last stage, children, especially boys, have a phallic focus. Boys desire their mother and see their father as a rival (Oedipus complex). This results in a fear of castration. Girls wish to remove their mother and connect with their father. Overcoming each stage without harm is of critical importance for structuring the basic dimensions of the personality. Otherwise, psychical energy remains stuck in the period where the harm took place and, consequently, trapped in whichever personality dimension is characteristic of that period. This wound means that the individual does not have enough energy to cope with the rest of their life. It is thought that everyone's basic personality is established by the age of five or six. The O. complex is overcome by a male child identifying with their father, or by a female child identifying with their mother. This is followed by a period of latency, which continues until adolescence, during which sexual interest is calmed. When children enter adolescence, the jumbled sexuality of childhood is replaced by a sexuality focused on the reproductive organs, which is the characteristic of adulthood.

Third argument: in order to survive, every individual must maintain some degree of psychical balance. When subconscious impulses, especially those driven by sexuality and aggression, clash too much with the values of the outside world and with the social values internalised by the individual, the individual develops defence mechanisms in order to maintain their psychical balance. Through a process that the individual is unaware of, these mechanisms hide the drives from the outside world and particularly from the individual themselves. These mechanisms always operate subconsciously, reining in this insidious anxiety enough to keep it at a tolerable level. If the individual is aware of all this, they can try to eliminate the inner fear evoked by these impulses by developing healthy conscious mechanisms such as avoidance or satisfying the drives in more acceptable ways. However, this process depends on whether the person is able to openly come to terms with their drives. As such a reckoning would throw most people unbearably off balance, subconscious mechanisms are developed, which the individual unknowingly operates. This comes at a heavy price. A large proportion of psychical energy is expended on resisting this unseen enemy and nameless inner turmoil, resulting in mental imbalances and disorders of varying severity. Of the numerous defence mechanisms, the most common are suppression (forcing things into the unconscious), denial (not admitting its existence), projection (attributing the problematic impulse to others), opposite reaction formation (e.g. showing excessive love in order to hide hate), rationalisation (e.g. "hitting someone for their own good"), intellectualisation (partially placating the emotion that provokes anxiety by extracting its affective load and turning it into a purely intellectual object, e.g. someone who has hidden their homosexuality engaging in academic research on the subject of homosexuality), displacement (getting angry at work and beating your spouse at home), concretisation (getting angry at work and punching a sandbag at home), fantasy (imagining winning the lottery), identification (emulating a hero figure that is free of the negative feelings), sublimation (e.g. directing aggression into a violent sport), partial gratification (stroking the head or cheek of the forbidden sexual object, patting them on the back). The only way to heal the mental discomfort or illness is to identify the drives behind the anxiety and consciously overcome them in ways that are acceptable to society and the individual. Freud first tried hypnosis to reach the subconscious. When hypnosis was inadequate, he developed various psychoanalytical tools: working with the client to interpret dreams, free associations, parapraxis, particular accidents, obsessive behaviours, phobias, jokes, meaningful forgetfulness, etc. Later, his followers added more tools, including games and reflective tests.

This is a rough summary of Freud's main arguments. Before assessing them from the point of view of historical materialism, a historical debt must be paid by those trying to develop a Marxist theory of the individual, and indeed by all others who have toyed with the subject of the individual; despite almost a century of innumerable, justified and unjustified, attacks on Sigmund Freud, he should be celebrated as the true pioneer of the theory of the individual. No serious theory of the individual could be constructed by undervaluing or completely ignoring Freud.

The most valuable aspect of Freud's first argument is that it was the first time in scientific history that the human psyche had been presented as a specific topic of research. It is easy to forget this in an era when such an approach goes without saying. Freud conceived this new scientific object as a system or structure that was largely unconscious. He was right on that point.

On the other hand, conceiving this structure as id, ego, and superego is problematic. This construct of Freud relates to his view that the individual psyche is directly connected to the instincts. The id is full of both pleasure-seeking feelings stemming from the life instinct and aggressive feelings arising from the death instinct. These feelings compel the individual to seek immediate gratification. The ego either restricts this gratification or places it in a framework that is acceptable to society. In my opinion, this is where the theory goes awry. The claim that there is a special psychical energy like "libido", separate from organismic energy, is ungrounded. Energy is an unequivocally concrete thing. Where could a psychical energy that is separate to physical energy be accommodated, bearing in mind that there is no such thing as a psyche separate to the body? As already stated when developing the historical materialist viewpoint, it is the psyche that shapes the whole body, not just the nervous system. It is correct that in this respect it operates like a system, but it does not have its own special domicile. Therefore, Freud's earlier topographic model, classifying the psychical states as unconscious, preconscious, and conscious, is more useful than this structural model.

The idea that the unconscious part of the psyche is completely dominated by the drives of sexual gratification and aggression is simply wrong. Even Freud's immediate successors in the psychoanalytical discipline rejected this view. Alfred Adler argued that the chief determinant of human behaviour was the search for superiority in order to overcome feelings of inferiority. Carl Jung argued that alongside the individual subconscious, a comprehensive subconscious common to all people was also a determinant. Erik Erikson believed that rather than sexual drives, the main determinant was the search for identity. Karen Horney proposed that the subconscious was more influenced by a myriad of problematic relationships creating distress in the child. Harry Stack Sullivan rooted the main problem in the lack of affective warmth in face-to-face relationships. In short, even the first Freudians rejected the idea that the human psyche was a slave to sexual feelings and aggressive drives. In the historical materialist viewpoint presented in this book, it is face-to-face relationships that are the main determinants of the human psyche, and these are shaped by culture. While it is true that the unconscious may host feeling-thoughts that could trigger psychical problems, it must also be remembered that it is basically comprised of individual habits that originated from society.

The *second of Freud's arguments* was the most problematic. According to this, from birth onwards every child's development is influenced by the different somatic regions of sexual sensibility in a predetermined order. By the age of five or six, they have acquired a psychical structure that will shape them throughout their entire life. While all the Freudians accept the special influence of the early period of development, they refute any idea that psychical development is frozen at such an early point. Erik Erikson's eight-step view of the psychical development process, lasting from birth to death, has been very influential in this area. Erikson equated the successful completion of each step to a specific societal relationship framework. (Erikson's stages were: infancy, learning to walk, early childhood, elementary school age, adolescence, young adulthood, adulthood, and old age.[57]) The focus of Freud's second argument is the O. complex, which supposes that boys aged between five and six sexually desire their mother and wish to supplant their father. When confronted by a powerful father, this desire causes the boy to feel both guilt and fear of castration. At this stage, the child internalises the rules of society, acquiring what Freud referred to as the "superego", and overcomes the O. complex by identifying with their father. It is through this mechanism that unbridled sexuality and aggression can be overcome or, in other words, a civilised existence can be established. Freud considered that this theory, based on the premise of an immutable human nature, was an adequate explanation of history and culture. The analysis in his famous work *Totem and Taboo* rested on this idea. One way of describing Freud's thought process is as follows: regardless of the historical period and cultural environment, the unchanging reality is that every human juvenile is the product of the union between a man and woman, and the product of this union wishes to displace their father in order to copulate with their mother.[58] It is this of all Freud's ideas that has been most subjected to attack, and rightly so. It is so unrealistic that it cannot even be considered as a scientific hypothesis. For Marxists, it is Freud's approach to history and culture that invites the most criticism. As Erich Fromm remarked, it bears all the hallmarks of a male-dominant viewpoint and is extremely derogatory to women, suggesting that the lack of a penis is their overriding determinant.[59] Essentially, it concludes that civilisation is largely the product of *men* who have successfully overcome their sexual impulses and aggression.

Freud's *third argument* made an invaluable contribution to psychology. He correctly observed that we unconsciously develop defence mechanisms that govern many of our behaviours, in response to unconscious feeling-thoughts that would threaten the psychical balance if they crossed into consciousness. It is not just feeling-thoughts that convey sexuality and aggression, but all kinds of feeling-thoughts disapproved of by society (and hence ourselves), that are hidden from the consciousness by these mechanisms. Freud and his followers provided an excellent description of these mechanisms. It is a psychological reality that we suppress the tendencies we dislike, even denying their existence when necessary. Projection, the act of attributing our flaws to others, is a widespread behaviour; the generally accepted view that those who shout loudest about lofty principles are mostly immoral and unprincipled is one of many examples. It is also true that we develop opposite reactions, such as behaving entirely contrary to our feelings in order to hide negative feeling-thoughts. With their analyses of these numerous defence mechanisms, Freud and his successors truly left a golden key to understanding the human psyche.

Moreover within the framework of this third argument, Freud's techniques for reaching the subconscious were extremely effective for reaching the unconscious parts of the psyche, for example the interpretation of dreams, free associations, parapraxis, obsessive behaviours, phobias, etc., or the games and projective tests developed later. Child psychologists successfully employ preconfigured games to identify psychical disturbance in children at an early age. In conclusion, leaving aside his obsession with sexuality and aggression, Freud made a vital contribution to psychology by developing procedural knowledge and theories on the operation of the system known as the psyche.

As long as the historical materialist viewpoint shuns this contribution, it will fail to explain how concrete mediations operate in the transition from social to individual. Every culture installs specific behaviour patterns in the individual through the medium of face-to-face relations. The dominant culture determines which things will be approved or disapproved. At the same time, many cultures carry conflicting values. Human juveniles assimilate these cultural behaviour patterns in their own unique way. As the life and concrete experience of every young individual are different, every psyche, every person is unique. The only way to analyse this individuality is by using the tools of psychology. Without using analytical tools like psychical balance, internal conflict, anxiety, and defence mechanisms, the behaviour of a concrete individual cannot be explained. The creator of these tools was none other than Sigmund Freud.

b) The Distinctive Trait Approach and Marxism

The distinctive trait approach was initially developed by Gordon Allport from the 1920s onwards and later further developed by Henry Murray and Raymond Cattell. It has been supported by a great deal of research, becoming the most widely used approach in personality psychology. Its main arguments can be summarised as follows.

First argument: the best way to describe personalities is to firstly determine the adjectives designated to them in language, then group similar terms together to reduce the overall number. Through a process of testing subjects, personalities can be placed on a particular rung on the ladder of each distinguishing characteristic. Allport identified over 4000 such words in the English language.[60] Murray later defined twenty-seven needs of psychological origin, considerably benefiting from his interest in psychoanalysis at the beginning of his professional life.[61] Cattell used the method of "factor analysis" to come up with sixteen basic characteristics that could be used to define personality.[62] Over time, he shortened the list using ever more advanced statistical techniques, eventually formulating the famous, widely accepted Big Five model of distinguishing traits. These surveys, which have been translated into numerous foreign languages, have achieved a high level of validity and consistency in environments where English is not the common language. Simply put, the Big Five model appears to be of universal value.

Second argument: according to proponents of the distinctive trait approach, these traits remain steadfast over time and in various situations. For example, someone who shows a high rate of extroversion will get a similar result in tests applied years after the first. The same person will also display a generally extrovert behaviour in different environments, such as work and home. The distinctive trait approach does not claim that everyone will behave in a completely consistent manner in every situation at any time; it argues that knowing a person's distinctive trait makes it possible to predict how they will generally behave.

What is the correct response to these arguments? Firstly, it is important to note that advocates of the distinctive trait approach are generally silent on the origins of the traits they have determined. There are those who claim they are evolutionary or are in some way related to heredity. But the most common attitude of the theory's proponents is to describe the traits as thoroughly as possible, while avoiding theoretical explanations. In fact, it is this more than anything else that has led to the notable success of the approach. Regardless of which school of psychology they follow or identify with, the vast majority of psychologists use these personality tests, and not just for academic research. These distinctive trait measuring tests are used in numerous fields of life from recruitment to promotion, from mental health assessment to choosing a career, from deciding which psychiatric technique to employ to managing social welfare schemes.

Having been applied to different clusters of people in various countries, at various times, and repeated at regular intervals, these tests have become something of a standard, and their application and evaluation is considered to be highly reliable. Consequently, they have become indispensable practical tools for all those working in the arena of personality theory, even those coming from a Marxist viewpoint. Although common agreement has not been reached on some aspects of these personality description tools, they are clearly extremely useful for descriptive and comparative purposes. One area of disagreement is the naming of each distinctive trait. As already mentioned in the second chapter (*see* section "The Description of the Soul"), the preferred set of names in their English language version is represented by the acronym OCEAN: these are Openness, Conscientiousness, Extroversion, Agreeableness, and Neuroticism. There is no complete agreement on the content of some of these dimensions.

Nevertheless, the content that is generally accepted can be summarised as follows: at one end of the openness dimension is imagination, readiness to various experiences, and independence of views, while the other end is dominated by narrow realism, mediocrity, and conformism; at one end of the conscientiousness dimension there is orderliness, caution, and sense of duty, and on the other end there is disorder, carelessness, and neglect of duty; at one end of the extroversion dimension sits openness to relationships, fun-loving, and social predisposition, and at the other end shyness, solemnity, and aloofness predominate; at one end of the agreeableness dimension are soft-heartedness, trust, and benevolence, while at the other end are hardness, scepticism towards others, and spurning cooperation; at one end of the neuroticism dimension are calmness, self-confidence, and being comfortable in your own skin, while at the other end are anxiety, lack of self-confidence, and apprehension.

The distinguishing traits that make up the Big Five are independent from each other. Each dimension is measured according to a scale, which might contain 100 points for example, and any of these points might be appropriate for each subject. As can be seen, with this approach there are no pre-defined overriding character moulds that subjects can be fitted into. The measuring process results in a different, individual character description for each subject.

c) The Biological Approach and Marxism

In the 1960s, when Hans Eysenck began to argue that personality was based in biology, most psychologists believed that children arrived in the world like an empty tableau, upon which their personality was drawn by the effects of their unique upbringing. These days, this view has radically changed. Most of today's psychologists believe that genetic factors, and therefore heredity, has an impact on personality. The chief arguments of those who believe that biology plays a major role in the formation of personality can be summarised as follows:

First Argument: like the researchers who used factor analysis to arrive at the Big Five distinctive traits, Eysenck also used factor analysis to extract the basic dimensions of personality. After many years of research, he concluded that there were three main determining characteristics for everyone: extroversion/introversion, "neuroticism", and "psychoticism".[63] Eysenck argued that these all originated from biological structure. The structure of extroverts has a high stimulation threshold, so they are predisposed to noise, crowded environments, loud music, and fast motion. In contrast, introverts are more comfortable in calm, quiet, and uncrowded spaces. His definition of "neuroticism" is identical in meaning to emotionality. People with a neurotic structure are more emotional and prone to anger and depression. "Psychoticism" is a form of malevolence. Psychotic types are "egocentric, aggressive, impersonal, cold, lacking in empathy, impulsive, lacking in concern for others, and generally unconcerned about the rights and welfare of other people".[64]

Second argument: genetic inheritance is the determinant of the main personality tendencies, but not of the specific details of the distinctive traits. Psychologists refer to these main tendencies as "temperament". There is no common agreement on the number of temperaments, with various people citing five, seven, or even nine types. Nevertheless, there is an overall agreement that most people match one of three temperaments, which are: outgoing, dynamic, emotional. Research among children revealed the following three prominent temperaments: compliant, unruly, shy. These groupings roughly correspond as follows: compliant children grow into outgoing adults; unruly children grow into dynamic adults; shy children grow into emotional adults. In parallel with life experiences, these main characteristics develop into one or other of the characteristic traits. However, as they are rooted in genetic structure, it is impossible for them to undergo a radical transformation at any time of life. In other words, after a certain point, the limitations of biological structure cannot be overcome.

Third argument: among those who argue that heredity plays a significant role in personality formation, some believe that the basic psychological mechanisms are evolutionary, i.e. that evolution determines our psychological structure in the same way that it determines our physical structure. Natural selection increases the probability that those with the most favourable genetic makeup survive and reproduce. Psychological characteristics should be thought of in this framework. For example, behaviour patterns advantageous to group solidarity mean that individuals are more supported and therefore are more likely to survive and reproduce.

Fourth argument: electrical activity in various parts of the brain can now be measured, especially using electroencephalography (EEG). The levels of different hormones can also be measured. There are also various studies aiming to research the connection between particular gene clusters and temperament. These technical advances make it possible to associate certain psychical states with biological structures or activities with a greater degree of reliability. Consequently, the possibilities for using the organism to reach psychical conclusions are gradually expanding. For example, drugs developed to counter depression are these days used with a high degree of success.

The important point to note when criticising arguments that hinge on the primary importance of biological factors to explain the psyche is this: Marxists cannot in principle object to the inclusion of biological elements. One of the main claims of all kinds of materialism, and consequently historical materialism, is that the psyche can only be found in the body. The Marxist objection to the biological approach is that biology is dealt with as an *ahistorical* factor, in the same way that Freud viewed a changeless biological drive, sexuality, as the main determinant of human history. If the drive in question had been something else, such as hunger, the argument would still be unacceptable from the perspective of historical materialism. Even if the determining biological feature had been sociability, it would still have been resoundingly rejected from the historical materialist viewpoint, because a theory of human nature that rests on common characteristics found in each individual's body is totally *incompatible* with a historicist stance. For Marxists, human nature is the whole of social relations. This whole is in a constant process of historical change. Let us imagine that a particular type of society stimulates sexuality, making it the chief determinant of most people's behaviour. Another society might provoke the hunger drive, and the same could be true of a society established from socially minded individuals. For Marxists, such a society, for example a communist one, can only be established as a result of historical development, and not because it is in keeping with human nature. Simply put, there is *no such thing as an immutable human nature*; humans are the only living creatures that are the architects of their own nature. Human nature is and always will be whatever we make it.[65] Undoubtedly, this process is not random. There is a compass showing the way, a compass that is the product of the history of an entire civilisation, with its ideals, principles, and aims, and which is even today constantly being refined. But this compass is the product of history, not immutable biological data.[66]

The view that the temperament, or basic psychical behaviour, is largely hereditary is one that can easily be adopted by historical materialism. Why should neuroticism or placidity not be hereditary, just like a myriad of physical features like figure, hair colour, or gait? But the *content* of a behaviour cannot be hereditary. It is always determined by inter-individual relations, or in other words culture. The form of a behaviour, not what it does but how it does it, must be to some extent hereditary. This is backed up by research. Of the three dimensions proposed by Eysenck – extroversion/introversion, "neuroticism", and "psychoticism"– the heredity of the first dimension has been most substantiated by research. Comprehensive studies, in particular with monozygotic and dizygotic twins, have made it possible to identify which attitudes stem from the surrounding environment and which are genetic.

The general agreement among Eysenck and the psychologists who share his views is that people are either outgoing/introvert, emotional/cold, or agreeable/disagreeable. It should be noticed that, aside from some differences in vocabulary, the ideas of Eysenck and the temperament psychologists are very close. Eysenck's extroversion/introversion dimension is more or less identical to the dimension they describe as "outgoing". The confusion begins when a characteristic is specified in one of the following ways: characteristics that are on the same dimension but at opposite extremes are presented as separate individual traits – for example, Eysenck's extrovert/introvert dimension; or a characteristic on one extreme of a dimension is specified and the reader is expected to infer that the opposite extreme exists – Eysenck adopts this approach for "neuroticism" and "psychoticism" (when talking of "neuroticism" he expects us to understand that it is a dimension with even-tempered at the opposite extreme; and when he talks of psychoticism, we are expected to infer that it is a dimension with agreeableness at the other extreme). The temperament psychologists also name a characteristic on one extreme of a dimension in order to specify the whole dimension, expecting the other one to be inferred. Consequently, while everyone says *almost the same thing*, the reader is left with the impression that they are looking at very different clusters. For instance, as the dimension named "outgoing" by the temperament psychologists has introversion at its opposite extreme, in reality it cannot be much different to Eysenck's extroversion/introversion dimension. The dimension that the temperament psychologists call "emotionality" has calmness at the opposite extreme and largely resembles Eysenck's "neuroticism" characteristic. Lastly, the temperament psychologists' dimension of sociability (agreeableness) has hostility to others at the other end and is therefore extremely similar to Eysenck's "psychoticism" characteristic. As seen, the confusing aspect is that Eysenck and other temperament psychologists use the name of characteristics located at *different extremes* to name the *same* personality dimension, such as "sociability" and "psychoticism". However, on examination of their concept definitions, it is clear that they mean virtually the same thing.

When these three personality dimensions are looked at from a hereditary point of view, it seems natural for dynamism-extroversion, reserve-introversion, emotionality-"neuroticism", calmness-placidity pairings to be hereditary. They are characteristics not dissimilar to hair colour or height. In contrast, it seems unfeasible that sociability-agreeableness and its opposite extreme of "psychoticism"-malevolence can be hereditary. This is because the first two pairings relate to *form*, while this last dimension relates to *content*. In keeping with the historical materialist perspective argued in this book, such content is always *historical*. Placing cooperation (or its opposite, hostility) in the framework of heredity stretches biologism to an extent that is unacceptable from a Marxist perspective.

An interesting point is that the attributes of the four main temperaments that have existed since Ancient Greece are *entirely* compatible with the viewpoint of historical materialism. According to the Ancient Greek explanation of the universe as constituted by four fundamental elements, there is a temperament to match each element: lively-sanguine (air), introverted-melancholic (earth), imbalanced-irritable (fire), indifferent-cool (water). These are exactly the two dimensions whose relationship to heredity I just argued was perfectly acceptable. The existence of four classes (or types) is related to the preference in Ancient Greece for presenting each of the same dimension's extremes as separate characteristics rather than the dual approaches mentioned above. "Air" is at the opposite extreme of "earth", and *both together*, they constitute Eysenck's dimension of extroversion/introversion. "Fire" is at the opposite extreme of "water" and together they correspond to Eysenck's "neuroticism", the dimension also named "emotionality" by the temperament psychologists. However, there is no Ancient Greek equivalent to the dimension referred to as "psychoticism" by Eysenck and "agreeableness" by other temperament psychologists. As can be seen, centuries-old observations support the historical materialist perspective more than some of today's widespread psychological theories. Heredity can influence characteristics like irritability, outwardness, etc., but whether humans are hostile or friendly towards each other cannot stem from heredity. These matters are informed not by genes but by the culture formed from all the relations that determine a society, particularly the mode of production.

How should historical materialists approach the view that psychical characteristics, just like physical ones, are determined by evolution? According to the Marxist theory of the individual/society developed in this book, the individual psyche is actually the individual form of society. Therefore, the psyche is a product of history rather than evolution. But in order to live up to its historical variability, the psyche needs to have the *capability* to experience certain feeling-thoughts. These capabilities are the previously outlined "feeltypes", which are located along six different dimensions and are named according to the characteristics of their poles: Like/Dislike, Trust/Fear, Approval/Disapproval, Joy/Sadness, Pleasure/Pain, Comprehension/Confusion. The *existence* of each of these potential capabilities can only be explained by evolution, but with the development of society, the qualities and content of these sensibilities changed. For example, without the evolutionary sensibility of repulsion that deters humans from trying poisonous foods, the feeling of repulsion experienced in reaction to certain human behaviour would also not be experienced. Although it is a specific evolutionary psychical sensibility that makes this reaction possible, it is entirely down to each culture to specify what will be considered repulsive and how this feeling will be manifested. Our psyches are determined not by evolution but by *history*, or in other words by *culture*.

This is supported by the fact that psychologists who see psychical characteristics as evolutionary try to explain some contradictory behaviours that are functional for societal life as products of pre-society evolution. For example, cooperation is functional because it strengthens collective life. But what of fighting? It is functional in selecting a mate and therefore in the reproduction of successful genes. Evidently, once it is accepted that both opposing qualities are functional, there is nothing that cannot be explained through evolution. But such a general observation does not really explain anything. When it comes to the psyche, evolution can only explain the existence of an affective capability, *a faculty for sensibility*. From thereon, the o*bject* of this sensibility and *how* it will be experienced is down to *history*, not evolution.

One argument of the biological determinists is that some physical and psychical conditions are related. This view, as long as it is *interpreted correctly*, far from contradicting the historical materialist approach, actually supports it. The Marxist theory I try to develop here states that the psyche is imprinted into the body. Therefore, identifying an organismic response to a psychical state is entirely in keeping with the materialist approach. For example, as already stated, it is highly likely that the temperament types described since the days of ancient Greece originate from gene clusters. Yet, it is only possible to explain individual behaviours through biological structure in rare situations when a certain physical characteristic has become so dominant that it becomes an active determinant. This might happen in the case of severe illness or disability. Apart from these exceptions, the general rule is that individual attitude and behaviour are determined by societal relations. In other words, the general rule is that individual relations are the active determinant. Nevertheless, it is entirely natural to find that an individual state arising from societal relations has a somatic manifestation. For example, an individual might experience anger or fear in a particular relationship. These emotions will have a number of somatic symptoms, for example high blood pressure or tightness of the muscles, but it does not mean that anger and fear are organismic rather than societal. Let us take the example of some biological symptoms of sexual arousal that have been revealed by recording the brain's electrical activity and various hormone levels. What does this show? It simply shows that psychical activities are reflected in our physiology. It is not evidence that this activity is of biological origins, although in fact it is possible to affect the psyche by dealing with physical symptoms, as in the case of drugs that improve positivity, increase morale, or relieve depression. But this facility does not detract from the fundamental reality that positivity, morale, and depression stem from individual relationships and are therefore determined by the surrounding culture. The fact that a person's pain can be reduced by giving a drug-free placebo[67] is clear proof that the main determinant is societal relations and that these are reflected in the body. In summary, just as the real determinant of a book is the meaning of its written content rather than its paper, ink, or font, it is the symbolic meaning of individual activities that is the real determinant of psychical state, not the activities of nerve cells, muscles, or hormones.

d) The Humanistic Approach and Marxism

The "humanistic approach" emerged in the 1960s, challenging both the Freudian approach of attributing human behaviour to the dominance of the id and the Behaviourist approach of focusing on the involuntary results of learning processes. Believing that these approaches overlooked individual free will and responsibility, this new psychological approach gradually garnered widespread support. The approach's "… roots lie primarily in two areas: existential philosophy which is decidedly European in flavour, and the work of some American psychologists, most notably Carl Rogers and Abraham Maslow".[68]

The main arguments of the approach can be summarised as follows:

First argument: the adult part of society generally consists of individuals who organise their lives according to free will and, consequently, feel personal responsibility for the things they do. Individuals are generally self-aware. Their most fundamental need is to develop themselves. The direction this takes is dictated not by society but by the different potential inside each person. Self-actualisation happens through fully developing this potential. People must be all they are capable of. Everyone must stay true to their own nature.

Second argument: according to Rogers, people who do not have the strength to accept themselves as they are have a distorted view of their own behaviour and even have a tendency to deny their feelings. This prevents them from fully developing their potential. Rather than providing the client with explanatory interpretations, the task of humanist psychiatry is to help individuals see themselves as they really are by creating an accepting environment with no preconditions and reminding them of their own words when necessary. The secrets that people bury inside prevent their ability to develop freely. The path to personal development is to disregard pressure from others, express feelings as they really are, and free your inner power to develop.

Third claim: according to Maslow, all people have common needs although the way they are met varies between cultures. One group of needs stems from deficiency motives, and another from growth needs. These needs are hierarchical and unless the lower need is to some extent satisfied, the next need up in the hierarchy cannot be met. At the very bottom are physiological needs like breathing and eating. Next is the need for safety, followed by the need for love and belonging, and then for respect. At the top is the need for self-actualisation.[69] Only a tiny minority of people can meet their need for self-actualisation. From time to time, these fulfilled people become totally absorbed by what they are doing or living, thereby reaching the peaks of human experience. The uplifting experiences enable them to know themselves better and increasingly satisfy their need for development.

The first comment on these arguments is that their theoretical value is *very low*. The claim that all people share a common volition for self-actualisation is a view that can only be observed in advanced capitalist conditions, and it is devoid of historic, and therefore scientific, substance. It is not hard to understand the allure of such arguments when living in the very kind of societal conditions described. In fact, exhortations on "good mental health" or "being yourself" are often found in the ranks of best-selling books aimed at the general public. But that is no guarantee that they have anything worthwhile to say.

Secondly, these arguments are self-contradictory. If everyone has an inner power that is waiting to be developed, the inner power must therefore be pre-determined; therefore, it is contradictory to argue that the individual can behave with free will. If it is a basic need to use a latent, potential will, it is dubious to claim that the will is entirely free and that people can choose how they behave.

Thirdly, these arguments are not original. The idea that people may distort or deny behaviour and feelings that conflict with their self-evaluation was proposed far prior to this by Freudian psychoanalysts. It is no different from the idea of "defence mechanisms". The notion that these mechanisms can be overcome by bringing them into the consciousness and that over time painful secrets fester in the mind is well-known in psychoanalysis. The psychiatric method is clear: enable consciousness to strengthen the ego and then try to adopt an appropriate rational behaviour in order to solve the problem.

Fourthly, this approach, particularly Maslow's hierarchy of needs, is based on an erroneous understanding of human nature, which is as widespread as it is wrong. This crude empiricism, commonly expressed in maxims like "A hungry bear never plays" or "bread first, then freedom", is entirely at odds with historical materialism. The view that, in all modes of production, all historical eras and all cultures, human needs are identical and ranked in the same hierarchy is *totally unacceptable*. The reality is that human needs are defined by the era and conditions in which they are experienced, in other words, by the culture. For example, if people living in one form of society can easily satisfy their hunger or easily enjoy sexual satisfaction, these cease to be "fundamental human needs" for that form of society. Aside from that, to assume that everyone arrives in the world with a "development need" and present this as the greatest path to humanisation is nothing more than pure fantasy. In summary, while the discourses of the humanistic psychologists might be appealing to the contemporary world, particularly the name they adopted for themselves, they are *totally lacking in substance* when examined from a theoretical point of view.

Marxists have nothing to learn from this approach, except for the diagnosis of some mental conditions created by advanced capitalism.

e) The Behaviourist/Social Learning Approach and Marxism
Behaviourism was pioneered by John B. Watson in 1913–1914. The second major name in the movement was Burrhus F. Skinner. In 1950–1960, traditional behaviourism was transformed by Julian Rotter, Albert Bandura, and others. The behaviourist arguments can be summarised as follows:

First argument: if psychology is a science, psychical states, such as "mood", that cannot be directly observed should be set aside, leaving just observable behaviours to be examined as a scientific object. Behaviourism is rooted in the work of Russian scholar Ivan Pavlov, who made his name at the end of the nineteenth century. Pavlov discovered the mechanism of conditioned reflexes by observing that when an artificial stimulus, like a bell, was repeatedly added to a real stimulus, like meat, eventually a dog would begin to salivate at the sound of the bell even if there was no meat. The behaviourists asserted that human and animal behaviour could be explained by conditioned reflexes. As well as Pavlov's conditioning mechanism, they also believed that the main key to understanding human behaviour was operant conditioning, which showed that the spontaneous behaviour of animals, humans, and any organism could be increased or decreased on one hand through reinforcement (approval, reward) and on the other through punishment (negative reinforcement). According to the behaviourists, people acquire a reaction pattern through a specific set of circumstances. They then undergo a *generalisation* process, through which they apply this same pattern in all similar situations. The character of every individual is built from these generalised reaction patterns (reflexes), which undergo new conditioning over time and can be intensified or abated through reinforcement or punishment. Watson believed that the entire character was formed in this way, stating: "Give me a dozen healthy infants… and I'll guarantee to take any one at random and train him to become any type of specialist I might select – doctor, lawyer, artist, merchant-chief and, yes, even beggar man and thief."[70]

Second argument: according to the second wave behaviourists, it was not just observable behaviours, but also inner events like thoughts and attitudes that could be subjected to conditioning. Rotter believed that an inner world conditioned in this way could acquire new perceptions, values, and most importantly *expectations*. This new social learning theory, which was added to the traditional view of behaviourism, suggested that human behaviour was determined according to the values attached to certain stimuli and conditioned expectations. According to Bandura's theory of reciprocal conditioning, external determinants like reward and punishment, together with inner determinants like beliefs, thoughts, and expectations, form a system based on reciprocal interaction. Ultimately, "…internal states, the environment, and behaviour all affect one another".[71]

Third argument: successful psychiatry depends on changing behaviour rather than raising the level of consciousness. Therefore, the aim is to change behavioural patterns with conditioning practices that will have a rapid effect. One method is to use the classic conditioning mechanism to create a systematic desensitisation to the source of the problem. Another method is to encourage positive behaviours by reinforcing some day-to-day behaviours at the same time as punishing others. A third method is to heighten client expectation of success, which employs methods such as reiterating past successes, learning from the experiences of others, or getting the client to take small steps with a high probability of success. Eventually, the person develops a belief that they are in control of their own behaviour and this belief, an "internal locus of control", works as a powerful lever in changing behaviour patterns.

How can these arguments be assessed from the perspective of historical materialism? On one hand, I think it is *generally true* to say that conditioning is the basic mechanism of learning. Although behaviourists do not give adequate credit to the characteristics passed down through heredity, it is true that after birth human juveniles are moulded through a process of continuous conditioning. Babies very quickly begin to associate factors accompanying pleasure stimuli as sources of pleasure and factors accompanying anxiety or pain stimuli as sources of stress or pain. The baby's expectations are created in parallel to these perceptions. At the same time, those around the baby approve or disapprove certain behaviours, providing a continuous operant conditioning. However, contrary to the ideas of traditional behaviourists, this does not equate to a crude shaping by external factors. As stated by Piaget, whose works are referred to above (*see* Chapter II: section "Feeltypes and Piaget's Theory of Soul Formation"), this learning process is an *active* one. The failure of the behaviourists to recognise that babies arrive in the world with a specific equipment, and that they are an active player in every situation, leads them to conclude that the character is formed by a passive process. This is where their mistake lies. In summary, the view that learning occurs through a series of conditioning mechanisms is correct, as long as it is remembered that it is an active process in which the object is at the same time the subject.

On the other hand, the behaviourists seem to approach conditioning as if it is an entirely independent variable. In other words, just like laboratory animals, humans are formed in whichever way they are directed by external forces. But from the perspective of historical materialism, these external factors are not randomly determined; on the contrary, they are factors validated by a particular historical era and society, i.e. a specific culture. It is true that conditioning exists. It is also true that conditioning is an active process that the subject also takes part in. But these truths are not enough to explain the individual. It should also be accepted that conditioning factors, methods, and aims are provided by the culture. Therefore, they are not random; they are the *product of a society* determined by the mode of production and all its other relations. It is only by taking all this into account that the conditioning mechanisms emphasised by the behaviourists enable us to understand a series of *mediations* in the establishment of the individual-society dialectical identity posited by historical materialism.

It is true that the second-wave behaviourists moved from seeing conditioning as a one-way mechanism towards emphasising the role of reciprocal interaction, and furthermore, as in the case of Bandura, saw that this reciprocal process would affect the whole system. But the proponents of behaviourism see the "system" as *any* inter-individual system consisting of behaviours with external and internal factors. Whereas, as just described, Marxists see *these systems as determined*. They are determined by history and society, in other words, the concrete culture in which the individual is living.

It must also be recognised that the behaviourists' approach to psychiatry is persuasive. Traditional psychology is essentially based on consciousness. The better the individual can see their subconscious, or more correctly, the better they can perceive their subconscious with an emotionally charged thought, the more the ego will strengthen and recuperate. But it is worth highlighting the inherent idealism here, i.e.: "The cure is knowledge." The behaviourist approach is much more materialist, i.e.: "The cure is to act." In this respect, the second approach is closer to historical materialism. Of the two principles, "If you want to change your behaviour, understand yourself!" or "If you want to understand yourself, change your behaviour", which is the more valid? When these two principles are weighed up properly, it is evident that compared with psychoanalysis, behaviourism is closer to Marxism.

f) The Cognitive Approach and Marxism

The cognitive approach first emerged in the mid-1950s and is still widespread today. George Kelly is seen as its pioneer, although he disliked the accolade.[72] In the mid-1970s, Albert Ellis became known for the psychiatric method he developed within the framework of the cognitive approach. This approach can be summarised as follows:

First argument: every individual has their own unique way of processing knowledge, and it is this that distinguishes us from one another. Everyone has a cognitive construct that consists of a series of evaluation dimensions with two extremes. For some, the measurements used to evaluate others might be criteria like "friendly/unfriendly", "happy/miserable", "well-dressed/badly-dressed". Others might use criteria like "cultured/uncultured", "attractive/unattractive", "trustworthy/untrustworthy", etc. As everyone sees the people and world around them differently, it is entirely natural that different people make *entirely conflicting* evaluations. Moreover, even if some evaluations used by two people overlap, their characterisations of the opposite extreme may not be consistent. For example, one person might consider the opposite pole of "extrovert" as "introvert", another might see it as "standoffish". In other words, everyone understands and judges those around them in *their own unique light.* When people see that their own evaluation criteria do not match their observations, they can partially change them. For example, if a person has a second encounter with someone they initially judged as "introverted" and realises that their expectation of "introversion" was wrong, they make appropriate changes to the evaluation criteria they use for introversion.

Second argument: using their own unique evaluation criteria, over time everyone forms fairly detailed evaluation constructs about the people they are close to. The information constructs that individuals form about their mother, father, siblings, close friends, etc., develop over long periods of time. These ready-made constructs are used whenever the behaviour of the relevant person is being evaluated. The data processing constructs may be loaded with positive or negative affectivity, depending on how past experiences are evaluated at any given time.

Third argument: from childhood onwards, people acquire a cognitive representation of themselves. Psychologists call this "self-concept". This perception is more than just an evaluation structure relevant to their true past and current personality. People have both "possible selves" and "ideal selves". In other words, people think there is a series of good or bad possibilities related to what might happen to them in the future. For example, "If I work hard, I could be a good teacher", or "If I don't deal with my jealous behaviour, it will ruin my marriage". As well as this, there is an "ideal self" that we develop through our own volition or through the influence of our environment. For example, "I am going to be a successful businessperson", or "I'm going to get into the national basketball team", etc. When we evaluate our own behaviour and decide on our next step, we use not just our existing self-concept, but also the information constructs related to our possible and ideal selves.

Fourth argument: understanding each other requires knowledge of each other's *evaluation constructs*. This knowledge includes the individual's own sense of self. If information constructs do not concur with reality, they lead to misunderstandings between people and also distort the individual's self-perception. If the constructs that the individual uses to process information about themselves and others are unrealistic, it causes inconsistency between behaviour, expectations, and results. The person gradually loses the ability to predict the consequences of their own actions and the behaviour of others, leading to anxiety. Cognitive psychiatrists aim to unveil the client's cognitive evaluation construct, enabling them to see its unrealistic aspects so they can modify them. A number of methods are used to achieve this, including tests, examination of drawings, and making videos.

How can the arguments of the cognitive approach be assessed from a historical materialist perspective? Firstly, it is important to note that the theoretical basis of this approach is extremely *shallow*. Even if the appropriate advanced research methods were used, it would still not go much further than presenting the notion: "Everyone sees the world through their own eyes." This general truth can only be of scientific value if it is backed by a theoretical framework. For example: where do these individual evaluation mechanisms originate? Are they chosen from a completely open-ended list? Or does each culture have its own restrictive list of options? How is this list compiled in different cultures? Which evaluation dimensions are more likely to be chosen in which types of culture or subculture? In which particular environments or relationships do people acquire their unique evaluation criteria? Etc. The cognitive approach can only contribute to personality theory if such questions are researched. If this were to be done, it could be of benefit to the historical materialist approach. *As it stands*, it cannot be said that the cognitive approach makes a worthwhile contribution to personality theory.

In other respects, focusing solely on the information processing aspect of the personality is a very restrictive and *extremely limited* approach. As already pointed out when describing the feeltype theory, apart from the Comprehension/Confusion dimension focused on by cognitive theorists, the human psyche also consists of other dimensions, i.e. Like/Dislike, Trust/Fear, Approval/Disapproval, Joy/Sadness, Pleasure/Pain. Any approach that does not include all these dimensions will inevitably fall short of the mark. Personality cannot be reduced to an information processing device, nor can its formation be simply down to acquiring right/wrong information. Affection, trust, respect, joy, and pleasure are also *indispensable* building blocks of the personality. It is inevitable that the theoretical side of any approach that ignores them will be inadequate.

Notes

1. As Murat Belge rightly pointed out, separating the emotional from the intellectual draws us towards "biologism" or bigoted "rationalism". Belge (1989, p. 275).
2. Charles Darwin stressed that what enables people to understand each other is the widespread similarity in the manifestation of emotions. Darwin (1872). This subject will be discussed in Chapter II, section "The Body's Impact".
3. Smith and Smith (1995, p. 112).
4. Darwin (1871, Vol. I, p. 101).
5. Darwin (1871, Vol. II, p. 390).
6. Darwin (1871, Vol. II, p. 393.)
7. Mahler et al. (1975, p. 3).
8. Vygotsky (1934, p. 228).
9. Adler (2013, pp. 72ff.).
10. Maslow calls these "peak-experiences". Maslow (1962, pp. 98–99).
11. Piaget explains this perception with the concept of "decentring" (a child no longer seeing itself at the centre of everything). See Piaget and Inhelder

(1969, pp. 94–95). This subject is touched upon later in Chapter II, section "Feeltypes and Piaget's Theory of Soul Formation".

12. "Becoming human is becoming individual, and we become individual under the guidance of cultural patterns, historically created systems of meaning in terms of which we give form, order, point, and direction to our lives." Geertz (1973, p. 52).
13. Yörükân (2006, p. 25).
14. Influenced by the World War I, Freud began developing his death drive theory in 1920. Over time, "the new dichotomy of life (Eros) and death instinct(s) takes the place of the original dichotomy between ego and sexual instincts." Fromm (1973, p. 442).
15. Batur (2007 p. 11)
16. Ibid. (pp. 120, 134).
17. Ibid. (p. 169).
18. Ibid. (p. 242).
19. Collins (2004, p. 172).
20. For two sources with excellent explanations of the fact that what constitutes the generic characteristic of humankind is that they themselves make up their own nature, *see* Mészáros (1970) and Ollman (1976). The first English publication of the latter was in 1971. It is interesting that both books, which in my opinion break new ground in understanding Marx, were first published in such close succession.
21. Darwin (1872).
22. This collective manifestation matrix does not just allow humans to understand each other, but also allows them to *deceive* each other. After all, the art of theatre is based on our common ability to attach meaning to each other's behaviour.
23. Similarly: Weeks (2010, p. 2). Also see. Foucault (1990).
24. http://www.physorg.com/news84720662.html. Accessed 26 June 2011.
25. Timur (2007, p. 153).
26. Durkheim (1914).
27. Sève (1969, p. 386).
28. Maslow (1962, pp. 136–137).
29. The sixth edition was published in 2006. The former emphasis on Piaget is missing from this latest edition of the handbook.
30. For an evaluation of Piaget in the context of the *Handbook of Child Psychology*, see Bond and Tryphon (1998).
31. Piaget (1970, p. 85).
32. Piaget and Inhelder (1969, p. 6).
33. Ibid. (pp. 9–12).
34. Ibid. (pp. 13–19).
35. Ibid. (pp. 53–54).

36. Ibid. (pp. 94–95).
37. Vygotsky (1934, p. 34).
38. Piaget (1952, p. 385).
39. http://www.marxists.org/archive/vygotsky/works/1931/higher-mental-functions.htm. Accessed 31 January 2013.
40. Vygotsky (1934, p. 6).
41. Ibid. (p. 34).
42. Ibid. (p. 44).
43. In Vygotsky's own striking words: "I am a social relation of me to myself." Holland and Lachicotte, Jr. (2007, p. 112, n. 7).
44. Vygotsky (1934, p. 107).
45. Ibid. (p. 201).
46. *See* Chapter III, section "Historical Materialism and George Herbert Mead"
47. Holland and Lachicotte, Jr. (2007, p. 105).
48. Wertsch (2007, pp. 178–179).
49. Holland and Lachicotte, Jr. (2007, p. 109).
50. Marx (1845, p. 29).
51. http://www.marxists.org/archive/vygotsky/works/1931/higher-mental-functions.htm. Accessed 31 January 2013.
52. The new updated edition published in 2004 was soon translated into Turkish.
53. Burger (2004, p. 4).
54. Ibid. (p. 5).
55. Ibid. (p. 11).
56. As stated earlier (Chapter II: section "The Structure of the Soul"), in Fact Freud's original impulse dichotomy was one between ego instincts and sexual instincts. He only began to base his theory on the life and death instinct dichotomy in 1920.
57. Erikson (1969, pp. 239–266).
58. Interestingly, the same idea appears in Denis Diderot's work *Le Neveu de Rameau (Rameau's Nephew)*. Diderot writes, "If the little Savage were left to himself...he'd wring his father's neck and sleep with his mother." Diderot (1762–1773).
59. "The libido was, according to Freud, male and there was no corresponding female libido. The woman was, in line with Freud's extreme patriarchal bias, not man's equal but a crippled, castrated male." Fromm (1973, p. 456).
60. Burger (2004, p. 165).
61. Ibid. (p. 173).
62. Ibid. (p. 175).
63. Ibid. (p. 247).
64. Ibid (p. 251).

65. Sève (2006, p. 87).
66. Ibid. (pp. 52–53).
67. Wagner and Silber (2004, p. 104).
68. Burger (2004, p. 304).
69. Ibid. (p. 314).
70. Ibid. (p. 375).
71. Ibid. (p. 388).
72. Ibid. (p. 444).

References

Adler, A. (2013). *Understanding Human Nature*. Routledge.

Batur, E. (2007). *Gövde'm [My Body]*. Sel.

Belge, M. (1989). *Marksist Estetik [Marxian Aesthetics]*. Birikim.

Bond, T. G., & Tryphon, A. (1998). *Piaget's Legacy as Reflected in the Handbook of Child Psychology*. http://www.piaget.org/news/docs/Bond-Tryphon-2007.pdf. Accessed 01 June 2011.

Burger, J. M. (2004). *Personality*. Thomson/Wadsworth.

Collins, R. (2004). *Interaction Ritual Chains*. Princeton University Press.

Darwin, C. (1871). *The Descent of Man, and Selection in Relation to Sex*. Two vol. John Murray. https://darwin-online.org.uk. Accessed 12 Feb 2011.

Darwin, C. (1872). *The Expression of the Emotions in Man and Animals*. John Murray. https://darwin-online.org.uk. Accessed 01 Sept 2011.

Diderot, D. Written Probably Between 1762–1773, It Was Published for the First Time in 1891, Over a Century After Diderot's Death in 1784. *Le Neveu de Rameau*. http://www.oeuvresouvertes.net/IMG/pdf/DIDEROT_NEVEU_DE_RAMEAU.pdf. Accessed 25 July 2011.

Durkheim, E. (1914). Le Dualisme de la Nature Humaine et ses Conditions Sociales. http://classiques.uqac.ca/classiques/Durkheim_emile/sc_soc_et_action/texte_4_15/dualisme_nature_humaine.pdf. Accessed 27 June 2010.

Erikson, E. H. (1969). *Childhood and Society*. Penguin Books.

Foucault, M. (1990). *The History of Sexuality*, Vol. 1, *An Introduction* (R. Hurley, Trans.). Vintage.

Fromm, E. (1973). *The Anatomy of Human Destructiveness*. Holt, Rinehart and Winston.

Geertz, C. (1973). *The Interpretation of Cultures*. Basic Books.

Holland, D., & Lachicotte, W., Jr. (2007). Vygotsky, Mead, and the New Sociocultural Studies of Identity. In H. Daniels, M. Cole, & J. V. Wertsch (Eds.), *The Cambridge Companion to Vygotsky* (pp. 101–135). Cambridge University Press.

http://www.physorg.com/news84720662.html. Accessed 26 June 2011.

Mahler, M. S., Pine, F., & Bergman, A. (1975). *The Psychological Birth of the Human Infant*. Karnac.

Marx, K. (1845). Theses on Feuerbach. In *Karl Marx and Frederick Engels Selected Works in One Volume* (pp. 28–30). Lawrence and Wishart, 1970.

Maslow, A. (1962). *Toward a Psychology of Being*. D. Van Nostrand Company, Inc.

Mészáros, I. (1970). *Marx's Theory of Alienation*. The Merlin Press Ltd.

Ollman, B. (1976). *Alienation, Marx's Conception of Man in Capitalist Society*. Cambridge University Press. (First published 1971).

Piaget, J. (1952). *The Origins of Intelligence in Children*. International Universities Press.

Piaget, J. (1970). *Psychologie et Epistémologie*. Denoël.

Piaget, J., & Inhelder, B. (1969). *The Psychology of the Child*. Basic Books.

Sève, L. (1969). *Marxisme et Théorie de la Personnalité*. Editions Sociales.

Sève, L. (2006). *Qu'est-ce que la Personnalité humaine?* La Dispute.

Smith, S. T., & Smith, K. D. (1995). Turkish Emotion Concepts: A Prototype Analysis. In J. A. Russell, J.-M. Fernandez-Dols, A. S. R. Manstead, & J. C. Wellenkamp (Eds.), *Everyday Conceptions of Emotion* (pp. 103–119). Kluwer Academic Publishers.

Timur, T. (2007). *Marksizm, İnsan ve Toplum [Marxism, Man and Society]*. Yordam Kitap.

Vygotsky, L. (1934). *Thought and Language*. (Newly Revised and Edited by A. Kozulin.). The MIT Press, 1986.

Wagner, H., & Silber, K. (2004). *Physiological Psychology*. Bios Scientific Publishers.

Weeks, J. (2010). *Sexuality*. Routledge.

Wertsch, J. V. (2007). Mediation. In H. Daniels, M. Cole, & J. V. Wertsch (Eds.), *The Cambridge Companion to Vygotsky* (pp. 178–192). Cambridge University Press.

Yörükân, T. (2006). *Alfred Adler, Sosyal Roller ve Kişilik [Alfred Adler, Societal Roles and Personality]*. Türkiye İş Bankası Kültür Yayınları.

Chapter III: Culture

Section A: Theory

1. Culture as Society

Culture is one of the most overladen terms used in the social sciences. Among many other things, it includes human behaviours, habits associated with eating, copulation, costumes/uniforms, jewellery/ornamentative styles, preferences, language, ways of speech, multiple art forms, architecture, social mores, roles, structures, institutions, production, exchange, circulation, distribution, consumption, leisure activities, sleeping habits, thoughts, knowledge, sciences, beliefs, religions, etiquette, jurisprudence, norms, social values, expectations, attitudes, judgements, foods, clothes, tools, materials, skills in numerous fields, education systems, behaviour models developed by different groups, symbols, communication tools and methods, political activities, legislative-executive-judicial institutions, ideological frameworks, techniques specific to every activity, and different fashions.

On careful examination of this open-ended list, it seems that the common function of all its features is to determine individual behaviour and mould it in certain ways. It is for this reason that culture is often defined as a mental system that regulates individual behaviour. As the individual is always determined through the medium of the brain, it may seem accurate to describe culture as a mental system. However, it must be remembered

C. Eroğul, *Marxism and the Individual*,
https://doi.org/10.1007/978-3-030-83662-7_4

that this system contains an infinite number of material elements. Indeed, defining culture only by its function is in effect cyclical: what defines human behaviour? Culture. What is culture? The societal elements that determine human behaviour.

All those who theorise over the concept of culture agree on one point: culture is passed on from one generation to the next. This leads some to describe culture as the habits of society or society's memory. The danger inherent in these interpretations is to think of society in the same way as one might think of an individual. It therefore seems more accurate to say *conventions* rather than habits and to conceive of memory as some kind of social record system. On the other hand, when looked at with regard to the individual, conventions are actually identical to individual *habits*. Their memory is imprinted into our bodies: "In habitual memory the past is, as it were, sedimented in the body."[1]

Culture is the most far-reaching concept in social science. Çiğdem Kâğıtçıbaşı writes that so far there have been over 160 definitions of culture, offering her own as follows: "Culture is the network of all material and behavioural arrangements acquired by a specific society so that its members can secure more satisfaction than can be found in nature; it includes all the knowledge, beliefs, art, morals, customs, skills and habits and social institutions developed by people as members of the society."[2]

The fundamental difficulty that confronts us with other sociological concepts is even more pronounced when it comes to culture. Scholars generally become trapped in a duality: on one hand, there is life and on the other there is society's *contribution* to it. This bias towards duality means that culture is generally understood as "non-biological", as if the following logic has been applied: humans are basically biological (a kind of animal), and culture consists of the things added to this biological state by societal life. Of course, the problem cannot be unravelled when it is posed in this way. The only scientific solution is to set this duality aside and see humans as *purely social* creatures.

Society is always a community of individuals. The individuals who make up a society are not simply biological creatures that are subjected to whichever qualities culture adds to them. Culture is in fact always within individuals. All the features of culture listed above, and all things related to human behaviour, are imprinted on the individual body. Those that are not are the material objects in which intellectual data are recorded with tools and materials suited to different purposes.

Individuals always move within relationship networks. Some of these networks are very tight and some are extremely loose. From rigid institutions to coincidental relations, they host a multitude of exchanges, from the most cohesive to the most flexible. But cohesiveness does not just stem from society's rules and sanctions. More importantly, it is *imprinted* on the individual psyche. This is true for the most powerful beliefs, norms, and attitudes of society. Society is always *inside* the individual.

The simplest way to conceptualise the *generic individual sociality* is to envision the daily life of any individual existing in a society. For one, every individual breathes, moves, eats, urinates and defecates, engages in sexual activity at times, and sleeps every night. These are biological necessities. With the exception of sexual activity, they are essential in order for the individual to survive as an organism. Sexuality is a requirement for the continuation of the species.

Some biological necessities are automatically met, such as breathing or sweating. However, it is not as if these are so unique to humans that they can be considered as definitive human features. It is only when they are touched by culture that even the most mundane phenomenon can turn into a human feature. For example, no animal uses their need to breathe in order to influence their own body or mental state. Yet this is exactly what yoga practitioners do. Breathing is biological, yoga is societal – in other words, an activity unique to humans. The same applies to sweating, movement, and eating. Many animal species sweat as well as humans, but only humans use deodorant to eradicate its odour. All animals move, but only humans engage in sport. Like humans, animals also eat, but what they eat is determined by biological evolution and imprinted onto their instincts. But when it comes to the human diet, it consists of the products of wild plants and animals that have been both domesticated and harvested by human hands, especially since the agricultural revolution. In the case of humans, their biological nutritional need is an *empty shell*, whose contents are determined by social relations. Some individuals are left hungry by society, left to scavenge from waste bins in order to satisfy their hunger, while others eat until they vomit. The eating habits of some societies has caused widespread obesity, but people elsewhere are emaciated with hunger. Even the functions of urination and excretion, which appear to be as biological as they come, are unique when it comes to humans. Unlike animals, humans do not defecate or urinate whenever they feel the need. They satisfy this need at particular times, in particular places, and in particular ways. All people sleep, but even this seemingly most natural need is met under conditions specified by society, i.e. particular times (e.g. during the day if you work a night shift), in particular spaces (e.g. in the bedroom, or on horseback), and on particular bedding (on bare earth if necessary or on an orthopaedic mattress). Sexual activity is the same. Animals also mate, but the sexual partners of humans are entirely determined by society. As a rule, people do not have sexual relations with their mothers, fathers, and siblings (with exceptions like Pharaonic families, which allowed marriage between siblings). That aside, who will pair with whom is determined by society. Marriages may or may not be arranged by families, inter-group marriage may or may not be permitted, partners are allowed from some groups but not others, while in some communities homosexuality is punishable by death yet in others it is accepted, etc. Human sexuality is almost limitless. With sexuality, as with all organismic needs, the element that actually represents the need is not an *empty drive*, it is a *particular* sexual *object*, which varies according to the society, environment, and person. As this is such an important point, it is worth emphasizing: our needs are defined by the *objects of gratification* towards which the needs are directed. The objects are always *societal*. To ignore this concrete reality by attempting to make abstract definitions of human needs is to fall into the trap of metaphysical thought. (*At the heart of this error is a belief that there is an unchanging human nature imprinted onto human biology.*)

Let us continue our mental exercise by again picturing the daily life of a miscellaneous individual in a miscellaneous society. We have observed that even the seemingly most biological needs of this individual are, in reality, social needs *delineated by objects of gratification*. Once we go beyond that, sociality becomes so apparent that explanation is redundant. When our imaginary individual awakes, they wash their face or have a shower. Maybe they brush their teeth. Then they get dressed. They might at the same time be listening to the radio or TV. They might put on make-up or perfume, depending on their usual habits. They have breakfast, possibly in a cafe if they live alone or perhaps in company if they live in a family unit. Then, depending on their age, they go to school or work. Their areas of work are determined by the social division of labour. Our individual's place in this division generally depends on class and status. Inside the workplace, they fulfil a role defined by society. After work, they go home, or perhaps to a bar, cinema, or some other leisure venue. In all of this, our individual conforms to the numerous rules and conventions determined by society. All the while, they greet others and join in relationships in ways that are specified by society. In all these relations, they use the language or languages taught to them by society, adopting the tone, expressive style, vocabulary, etc., that society finds acceptable in the given situation. When they get home, they watch TV, browse the Internet, read books, or play games – in other words, they fill their time with the possibilities offered by society. They go to bed alone or with another person (their partner if they have one, or someone they have met through the available social possibilities, or someone they have paid for a service). When they get up the following day, their entire life experience is once more shaped by their position in the *network of societal relations.*

Individuals are never completely passive actors in the roles they play. They use their own energy (even if it is conditioned) to enact numerous roles through their own volition. Depending on the society, some of these roles may sometimes be pretty much unavoidable. Yet, whether or not they take on various other roles is down to their own will. The individual is always party to random relations outside of their roles. In all these roles and relations, the individual to some degree conforms to convention and to some degree makes changes. When conditions allow, the individuals themselves create new roles. In every role they play, the individual manifests their personality, which is the specific product of their own past, i.e. of their past social relations.

Every individual starts out in life with some potential capabilities passed on through heredity. These innate propensities create a predisposition towards some social relations and a reticence towards others. The face-to-face relations experienced by a young juvenile may strengthen or weaken any of these tendencies. From birth onwards, human juveniles never surrender to the surrounding conditions in a completely passive manner. They are always active. The surrounding environment determines how effective and efficient their activities will be. The stronger a child becomes, the more possibilities it has to shape its environment according to its own preferences. During the growth stage, children are unavoidably subject to the rules of their own organism. For example, they cannot choose to leap forward into adolescence no matter how much they or anyone else might want it. But how the child experiences adolescence is determined by both society and the child's own personality as far as it has formed up until that point.

Individuals who are able to develop a strong ego while growing up can reject the roles allotted to them by society, fulfilling them in their own way to varying degrees and even creating new roles if conditions permit. If the social conditions are amenable and enough other individuals can be brought onside, this can bring about lasting changes to the social order. What individuals do is not *imposed by* society. Society only presents *possibilities* or *challenges*. Some individual preferences might prove extremely difficult at different times, but even the most challenging possibilities can be compelled into being by a sufficiently determined individual. It should be remembered that a society per se cannot sustain or change itself. This change is always down to the *individuals* within it. Consequently, even the survival of a society is always dependent on the activities of its individuals. Social transformation requires even stronger and more effective individual actions.

In summary, every individual psyche is formed by the concrete face-to-face relations inside the relevant society. Every psyche is an active participant of the society within which it exists. On the other hand, every psyche is imprinted onto a particular body. Therefore, on a superficial level it seems as if inter-psyche relations occur between bodies. Every culture provides each psyche with a multitude of intellectual and material possibilities. It makes some of these possibilities easy to take up and some more difficult. Every psyche mobilises different kinds of possibilities to differing degrees, depending on their own personality tendencies and the strength of their own ego. The untrained eye mostly observes moving bodies in these activities. But it is the psyche, i.e. the active element of culture, that moves. Ultimately, a culture is sustained, renewed, and changed thanks to its active elements, which are the determinants of its formation. Therefore, culture continually creates new active elements (psyches). This is its function.

2. *The Location of Cultures*

There are two principle cultural carriers: (1) individuals; (2) all living and non-living entities outside of the human race and all material and spiritual objects accumulated throughout history.

Let us begin with the second of these. As it has often been overlooked, it might be useful to start by looking at nature as a cultural carrier. As already stated, today's natural world with its flora and fauna is largely a human product. Even before the Neolithic age, before we even get to relatively advanced *Homo* species, humans had destroyed around fifty species of large herbivores through over hunting.[3] In contrast to the animals that we have domesticated to meet our various needs (especially dogs), nearly all animal species that have remained in the wild are in danger of extinction. Even the animals we domesticated have not been left untouched. For thousands of years leading up to the cloning process that is possible today, we have played with the biology of our domestic animals.[4] We did the same with plants. From burning forests to make way for fields to covering all arable land with our domesticated wild seeds, we have left our mark on the entire plant kingdom, culminating in today's genetic modifications. What we now refer to as *nature* is largely a product of culture.

There are countless cultural objects accumulated throughout history and pre-history, from our own remains and those of other creatures (fossils, bones, etc.) to the ruins of our ancient settlements and from stone age remains going back two and a half million years to a multitude of different implements filling the world's museums. Let us add to that the weapons we have accumulated from the beginning of the human journey, from the bow and arrow to the atom bomb. From pictures painted on the walls of caves to excavated instruments, these numerous art forms are another product of human culture. Since ancient times, we have worshipped various supernatural forces, over time developing advanced religions, creating rituals and types of worship, and attributing holiness or evil to a host of objects. Thousands of years ago, we created writing, resulting in a huge range of written forms, some of which we still do not understand. The objects carrying this writing are outstanding examples of our cultural existence. As writing advanced, we wrote books, developed printing techniques, and established libraries of tens of thousands of books. By the time we reached the stage of digital records, our potential to accumulate information became almost boundless, and we popularised the devices that conveyed this new accumulation mechanism as widely as possible. Perhaps most importantly of all, we created languages, thus launching *the most powerful tool* conceivable for forming, maintaining, and advancing a culture. In short, from the moment we emerged on the earth as humans, we have been carving the cultural phenomenon that defines our species into everything around us, be it material or spiritual. With all this accumulation, from the earliest times to the present day, a world emerged that is every *inch a carrier of culture*.

However, the main cultural carriers are *humans themselves*. Our bodies and our psyches are generically cultural objects. The human body is not just any animal body. It has been shaped by evolution into a body that can only accommodate a human. Natural selection is guided by the needs of cultural progress, and we are certainly indebted to that when it comes to our hands and brains. Furthermore, the influence of culture has not lessened with time. In fact, it has increased. Our biology and genetic make-up continue to carry as much culture as they can manage. The human body accommodates the most incredible artists, musical geniuses, painters, sculptors, and almost superhuman athletes. As already stated, culture is widely accumulated in objects, but this very accumulation also inevitably equips our bodies with the ability to manage it. Our psyches, which are the individual form of our society and culture, can only exist by shaping the bodies that host them into amenable forms.

An individual's genetic makeup is passed on to their descendants in varying degrees depending on the characteristic concerned. Similarly, individual behaviour patterns – roles – and the structures that build the relationship matrix of these roles continue their influence when the individuals that perform them leave their place to someone else. Hence, the patterns of one generation more or less shape the next generation. Some genetic characteristics have a strong impact on progeny, as do particular roles and patterns. Social behaviour and relationship patterns that have become potent enough to be embodied by institutions wield a powerful influence on the shaping of the next generation.

As well as all the external elements that convey culture, each individual body is also a cultural carrier. Individuals do not enter into a society as empty vessels. Right from the start of the species, humans have come into the world as individuals *determined by* society. This is more than just a matter of skin colour or eye shape. Every individual carries a multitude of societal skills, such as walking on two legs, using their hands for a variety of different purposes and the ability to speak. In short, they carry the history of the species in their genetic makeup, nerve cells, and endocrine glands. And it does not stop there. Every individual carries in their body their own psyche or character, which has been shaped by living in a society. We should not be misled by the word "carry". Perhaps it is more accurate to say that the psyche is "imprinted" onto the body. As the psyche is a product of sociality, social relations are embedded in the individual. Social relations, together with the psyche that is imprinted onto the individual body, also evolve and change. As the psyche is the individual form of society and the body is the carrier of the psyche, it should not be surprising that sociality is imprinted onto the individual body. (There are some striking examples of sociality being imprinted onto the individual body, such as people who have developed a hunchback through a life of carrying heavy loads, dentists with round shoulders due to their working conditions, people who have become asthmatic through air pollution, people who have become stooped from continually working in cramped conditions or who have become stiff from continually tensing their body.)

3. The Sociality of the Individual and the Individuality of Society: Mediations

To concretise the third level of dialectical identicality between the individual and society, there are two prerequisites: one is to show that every element of the individual is formed, developed, and changed by society; the other is to show that every element of society is formed, developed, and changed by the individual. This will demonstrate that the individual can only be formed through social means and that society can only be established through individual means.

The individual is a structure. As already pointed out, the active element in the structure is the psyche, and its passive elements are: (a) the body; (b) genetic build; (c) the influences encountered in the gestation period; (d) innate physical and intellectual capabilities and limitations; (e) language as the shaper of intellect; (f) innate character traits; (g) the role allocated to the individual from birth onwards; (h) the behaviour patterns acquired by the individual during the growth process; and (i) the character system that carries a lasting tendency.

The psyche is an active element. It uses energy. It cannot be reduced to simply a compound of passive elements. It is active, actuating, and transformative, even possessing the power to reshape its own passive elements. (For example, the body can be developed through regular exercise.) It is a system composed of a feeling-thought combination. Structurally, it is composed of the id, ego, and superego. The id and superego are largely unconscious, as is the ego initially. However, the ego can reach a high level of consciousness if it is enabled by the surrounding culture.

The human body and genetic makeup are a product of species evolution. The predominant factors of this evolution are the pressure to adapt to the conditions of the social order (including the natural environment) and deal with its challenges. When two opposite sex individuals of the same species copulate, a new distinct individual emerges, carrying a unique blend of the mother and father's genetic characteristics. This individual enters the world as a specific individual who carries the general characteristics of the species, elements of the mother and father's original genetic makeup, and the traces of events that occurred during the gestation period. Therefore, from the minute an individual is born it is unique. The key to sociality is this uniqueness. Indeed, it is the *uniqueness* of its members that distinguishes society from a herd.

As a species, all individuals are identical to each other; but, as individuals, they are different from each other. Just as establishing value identicality in an exchange depends on the disparity of the use value of the objects that are to be exchanged (i.e. they must be useful in different ways and if there is no difference then there is no exchange), in the same way, society can only exist if it is formed from *different* individuals. *Associative relations can only be established on the basis of mutual difference.* A community made up of identical individuals could not be a "society". As we can see, ants, bees, etc., can only constitute an animal community. *Society is unique to humans.* It is true that in animal communities, for instance in a beehive, there are sub-groups that are not identical to others because they perform different functions; but in these communities the segregation is not on an individual level. It is only in communities unique to the human species, in "society", that the segregation is on an individual level. Every individual is a combination of unique characteristics. They have a particular innate intellectual and biological potential. Each individual has their own unique specificities in every area, their own limitations and superiorities compared to the average of the species. Every individual is born into a language environment and a society's system of upbringing. This upbringing is culture specific. Practices such as suckling, swaddling, putting to sleep, washing, vaccination, etc., are defined by the culture. Nevertheless, the young individual never passively adopts the role allotted to them by society. From the moment a baby is born, it to some degree adapts this role to itself under the influence of its own specificities. Using the cultural means available, a juvenile gives its *own colour* to the role. The psyche is shaped through this active exchange, gradually acquiring a certain character. While this character is to some degree pliable and open to development, it must also enable the systemic underpinning that is a condition of existence for the psyche. Consequently, the character that emerges is both rigid and pliable. It is a product of the face-to-face relations established by the individual with their environment. It is both social and also specific to the individual. It is the product of sociality's *uniquely* individual way of life.

This character also *shapes* the individual body. The psyche establishes its existence through this shaping, which is basically written into the neural system, the hormones, and the configuration of the genetic makeup. Although it is largely internal, it is also reflected on the *outer* body.

Sensations, perceptions, feelings, stance, gait, stride, and scent, the way each part of the body is groomed, make-up, tattoos, clothing, jewellery – in other words, all the features that make each body unique – are vehicles through which the psyche (which is also unique) exists.

For the individual, society is primarily the other individuals with which they have face-to-face relations throughout their life. While the number of these relations varies from one individual to another, it is at maximum around 100 to 150 people. The psyche of every individual forms, develops, and changes within the process of long-term associations with this small number of people. Therefore, in order to analyse the individuality/sociality mediations of any society, we first need to shed light on the face-to-face relationship groups specific to the society, the roles they entail, the formations of the feeltype that determines the manner of these roles, and the material and intellectual possibilities presented to the holders of different roles.

Although their concrete shape changes according to the society, there are two types of face-to-face relationship groups that are essential to every society: one is the group that enables the formation of new individuals (psyches), and the other is the production groups that are a condition for the continuation of the society. No society can survive without renewing its individual members and continually providing the objects that are indispensable for material life. In other words, the organising of the family and of subsistence are the two main forms of face-to-face relations in all societies. A society in which these organisational forms were disengaged from face-to-face relations would be one where babies were formed in test tubes and raised in laboratories and all production was carried out by robots. As the likelihood of this type of society is fairly remote at the moment, we can leave it aside for now.

Apart from these two types of face-to-face relations that are certain to exist in every society, there are many other kinds of tight groups, which vary depending on the society-culture. Some of the most well-known are: relations that depend on totemic associations in tribal (ancestral, clan) societies; groups that are largely based on belief in a common religion; common festivities-celebrations that have become traditional although their religious roots may be forgotten; age groups; settlement groups; kinship groups; research-education-based associations; tightly woven profit groups; political or ideological groupings; artistic associations; sports associations; groupings based on a shared gratification (drinking friends, bridge, chess groups, etc.); groups based on strong affection (lovers, friends); and criminal gangs. Clearly, it is not just groups based on subsistence that vary significantly according to the mode of production and its stage of development, but also those based on family, kinship, love, or friendship. Therefore, in order to analyse the individual (i.e. society), we must start by learning the type and qualities of the face-to-face relation groups that exist in the relevant society.

The way these relationship groups function is more or less conventionalised according to the culture or sub-culture. From the most conventional institutions to increasingly flexible relationship patterns, including even random relations whose transience means they leave little trace, all societal exchanges are carried out by individuals. However, they generally do so by enacting certain roles, i.e. by carrying society within them.

Despite the fact that these roles are the same, *the way* they are performed varies according to the individual psyche. As mentioned earlier, "The Big Five" system developed to describe personality can be used to classify how these roles are played. This results in a series of depictions ("unaffectionate father", "calm boss", "honest manager", "irritable teacher", etc.). As these depictions depend on individual qualities, we need to examine how these qualities developed. The way to do this at the individual level is to look at *biographical stories.* Every individual grows up and finds a place in society in conditions that are unique to them. During the process, every individual (every psyche) is party to exchanges that influence, develop, gratify, stunt, or cripple the six basic feeltypes. In concrete, this is how everyone's biography is written. *Societal life happens in the form of these different individual biographical stories.*

Rather than looking at the issue of the participation of new individuals in structures, the correct methodological approach is to find out which structures are more effective, through which ways their effects are felt, and the impact they have on other structures. Actually, the issue of new individuals participating in society (the renowned topic of "socialisation") is a misleading one if it is put at the centre of everything,[5] because an individual is part of society, even before they are born. At birth, the individual is not an indeterminate "blank canvas", they are a being equipped with human potential. When an individual becomes an active element inside a social structure, they are simply changing position inside the entirety of social relations. From a state of being relatively passive, they transform into a more active protagonist. The importance of this transformation cannot be underplayed, but it would be wrong to think that the active element is a "brand new" element, completely alien to the social structure that has been joined. The socialisation of the individual and the individuation of society are not two alien processes, but essentially two faces of the *same process*. Hence, the participation of new individuals in society does not pose a real problem, which is why it is so misleading to become absorbed by such issues.

Structures shift through the practices of the individuals that play a role within them. These actions are to some extent determined by the behaviour pattern that society has assigned to that role. However, this determination has a degree of flexibility dependent on the role itself. An adult plays out their various societal roles according to their own character. In some situations, individuals may be able to enact their roles very freely. Moreover, they might completely step out of these roles. Just as such changes in individual roles may occur as a result of the same individual changing their behaviour due to a different active determinant, they may also occur as a result of an existing role being undertaken by a new individual who is determined by different influences. This is how structural transformation takes place. As a result, the structure concerned may impact other structures in a completely different way.

Every role presents the role holder with a series of material and psychical possibilities, which vary according to the society. Using these possibilities, the individual firstly mobilises their immediate circle of face-to-face relations. As this circle of face-to-face relations widens, new faces begin to enter it. If the individual has been able to grasp the intellectual-affective tools required by the society at that conjuncture, it may lead to significant transformations in the society. Thus, society finds the way to change itself through individual endeavours.

Social structures consist of all living individuals and the mass of material and intellectual wealth accumulated by society from past to present. As already stated, this accumulation is embedded in a wide variety of carriers from the genetic makeup of the species to a multitude of cultural objects, including the so-called "natural" environment. Therefore, the future of society to some extent *already exists today*. The existing material and intellectual accumulation are passive determinants of this future. Its active determinants are the ideas about the future projected by the individuals that make up society today. In the process of approaching the future, the new challenges faced by society will be met by new solutions created by these active determinants inside the framework of passive determinants.

4. From Individual Habits to Social Customs

Our daily practices, activities, and behaviours, or in other words human conduct, originate from three sources: (1) habits; (2) roles-structures-institutions; and (3) individual preferences.

Institutions, structures, and roles are the hardened-crystallised forms of social conventions. Their equivalents in the psyche are psychical formations with varying degrees of inflexibility. The most obvious example is unconscious practices. Just as skills like riding a bike, skiing, or even simply the ability to stand on two legs is imprinted onto our muscles and neural system, our habits too are nothing more than particular configurations of our body. Just like purely somatic mechanisms, these too are located in the neural networks that carry our psyche. As noted, the institutions, structures, and roles created by social convention take form on the psychical plane as specific habits. Just as institutions, structures, and roles mould societal relations into certain patterns, habits also keep our behaviours within more or less fixed patterns. Therefore, habits can be defined as individuated institutions, structures, and roles.

Society is a bundled mass of structures. Structures are the product of individual roles. Structures (and the roles that make them) have varying degrees of transience and permanence. The most permanent are those that have become institutions. Institutions themselves are differentiated according to their degree of permanence. Every society is made up of countless structures, some of which have fortress-like resilience and longevity and some that easily disintegrate. To analyse any society, its most robust, permanent, and influential structures must be traced.

One of the most prominent determining factors in any structure is the social area in which it is active. In historical materialist theory, areas of societal activity are divided into three main types: production, politics, and ideology. Like individual activity areas, these are also established through inter-individual relations. However, unlike the close face-to-face relations established by individuals through friendship, leisure, or sporting activities, these activity areas are far more important with regard to the reproduction, development, and transformation of society.

From the perspective of historical materialism, in order to unravel the mystery of a particular society, it is firstly necessary to look at how it (re-) produces the material means of life. It is crucial to realise that the question of how material life is reproduced is not a purely technical one. This question relates directly to the relations of production. The answer is primarily found in the ownership structure of the means of production. The reason is clear: for production to happen, in other words for a society to sustain its existence, workers (the labour force) and the means of production *must come together*. For this reason, whoever holds the means of production determines the characteristics of production and also which policies will be dominant in the major areas of society that lie outside of production.

The habits that have shaped and continue to shape the psyche are of course not set in stone. Their ability to change depends on the flexibility of the configurations they have already been subjected to. Some habits are extremely inflexible. For example, an adult would not stop walking and begin to crawl again, nor would someone who is used to eating with cutlery suddenly change to eating with their hands or tearing off pieces of meat with their teeth. Just like powerful societal institutions, these types of habits form foundations that are difficult for the psyche to modify.

Habits and roles-structures mostly change unconsciously. Without being aware of it, the individual acquires new habits and structures, developed by society, together with the accompanying roles. As society progresses, individual preferences also begin to have an impact on changes in societal relations. Individual preferences are conscious and deliberate. Individuals knowingly and willingly change their habits. They do the same when fulfilling their roles. Moreover, they spurn certain roles and try to develop new ones, triggering changes to social structures. The emergence of conscious preference is undoubtedly connected to a certain level of development reached by society. A society at a level that allows the psyche to make choices, at the same time presents the psyche with an array of sociocultural possibilities from which to choose.

The more flexible habits are, the easier it is to change them through individual effort and consciousness. These changes also have a transformative effect on the society's institutions, structures, and roles. Individuals can transform a society's institutions, structures, and behaviours according to the power of their social position or by dint of sheer numbers. It is not hard to grasp society–individual interaction in this framework. Institutions, structures, and roles instil certain habits in individuals. Individuals reshape institutions, structures, and roles using their status, individual power, or through strength of numbers.

According to Nicholas Rose,[6] when a society is confronted by a problem it needs to tackle, it builds new roles that will create individuals capable of overcoming it. The basic function of these roles is to create individuals that will be willing to solve the problem. How does this process work?

My explanation is that as the problem becomes apparent, a number of ideas begin to circulate (diagnoses, interpretations, suggested solutions). In addition to playing their existing roles, individuals who sympathise with these ideas adopt new behaviours that fit with the ideas they embrace, gradually leading to the emergence of new roles. For example, the early 1960s saw the first widespread emergence of socialist ideas in Turkey. Consequently, a series of new roles formed among those who identified with these ideas: member of a revolutionary party, revolutionary trade unionist, revolutionary journalist, revolutionary artist, illegal revolutionary, etc. In reaction to this, corresponding counter-roles developed, for example counter-revolutionary party member, counter-revolutionary, etc.

The emergence of these numerous new roles needs to be fully analysed. All the while they are only thoughts or cultural elements, they cannot be expected to impact on society. In order for such an impact to materialise, the roles capable of activating these elements must be formed. These roles can only be formed by those who identify with the ideas. Therefore, the individuals concerned must adopt *behaviours* suited to their favoured ideas on a fairly consistent basis. The formation of roles is a process, but it is not inevitable that these processes will crystallise into solid patterns. What mostly happens is a relative solidifying or crystallising in behaviours. When the social conditions change, these patterns unravel and new ones begin to emerge. As Rose stated, throughout their life every individual performs a variety of very different synchronic and diachronic roles. It is this continual individual dynamism that sustains, perpetuates, and transforms society.

5. From Culture to Feelings: Feeltypes

The feeltype theory is of great assistance when examining the mechanisms through which cultures shape individuals. Using this theory, we can systematically ask *the same questions* of different cultures. (This also allows us to *compare* cultures in a scientific way.) Some concrete questions for examining the psychical aspects of culture might be as follows; in the framework of the culture: who are the objects of affection? Which types of affection are valid for which objects? From what do people derive joy? How do people experience joy? Which situations garner the approval of others? Whose approval has more value? Which situations are humiliating? How is the humiliated person expected to behave? Which relationships provide security? What are the main sources of fear? Which relationships between whom are the source of pleasure? Which relationships cause pain? How do people satisfy their need for comprehension, in which environments and through which relationships? Which types of love, and with whom, are approved or disapproved? Which love gives pleasure and which gives pain? Are the relations that satisfy the positive extremes of the axes compatible or incompatible with each other? Does a relationship that satisfies one dimension conflict with another feeltype? The list goes on. As seen, firstly every feeltype axis needs to be examined in its own right with regard to the relationship between its extremes; then, the entireties and extremes of every axis must be investigated in relation to the entireties and extremes of every other axis. (Bearing in mind the psyche's need to ensure integrity, inner congruence, and exchange energy, this also makes it possible to evaluate the functionalities of different cultures with regard to *psychical health*.)

Every individual is formed and differentiated in some way by the culture they belong to. The hereditary characteristics of each individual are different, but more importantly, even individuals in the same subcultural groups have their own unique life experience. Every differentiation in experiences determines the psyche accordingly. The psyche subsequently experiences life with this newly determined form. This is what makes every individual *unique*. Knowledge of the feeltypes helps to understand the formation of individual characteristics, the integrity of the psyche, its inner congruence, and how it achieves the function of energy exchange. From the minute individuals are born until the day they die, they acquire personality characteristics that change, expire, or are replaced by others. Initially, the process is fluid and fast, but it quickly crystallises, losing its flexibility as time passes.

Every culture casts all relations, including face-to-face, into more or less set patterns (roles, structures, institutions). Everything outside of these patterns, including all random relations and even individual practices outside of social relations, is ultimately to some degree determined by the cultural environment. Also, each culture provides all the material and psychical materials available for all individual relations, conducts, or behavioural patterns. These tools are specific passive determinants. The way to analyse a particular culture is to examine the individual relations, conducts, or behavioural patterns with which the six basic feeltypes are established (structured, concretised). These concretisations are differentiated according to the society's classes, groups, and subcultures. For individual psyches located in these sub-formations, how a feeltype is experienced, to what intensity and degree, how comprehensive it is, whether or not it is integral, and whether or not the inner contradictions are reconcilable are of vital importance. By examining these, we can research the personality types developed by a particular culture or subculture. (What Abram Kardiner, following on from Erich Fromm, refers to as "basic personality".[7]) We can also assess how positive or negative the culture in question is from the point of view of psychical health.

Clearly, this is of great importance for individual development and, when necessary, for therapy. The relations that impact on the characteristics most important to psychical health must be developed or ceased, depending on the circumstances. The more aware an individual is of the situation, the more they can try to make changes to their life, either through their own willpower or with help from others. The more successful such efforts are for increasing individual happiness, the greater will be the positive impact on the whole society.

The more primitive the culture, the more identical are the id and superego. An individual who is just a cell of the society will directly identify with the society. This is only possible in very small, primitive communities, which are largely static and unchanging. It is inevitable that the individual psyche's positivity/negativity criteria will separate from societal expectations when the community grows, the layers become more multi-fold, environmental conditions diversify, and division of labour emerges. This development becomes unavoidable as just one societal expectation is replaced by many.

An individual is born into a certain cultural framework, where their concrete experiences provide them with satisfaction/dissatisfaction. In response to this, they develop appropriate reaction habits. Part of these unconscious habits consists of defence mechanisms developed to deal with negative events. As previously described, these mechanisms can take many different forms, such as repressing, resisting, denying, or displacing negative feelings, reducing their intensity, or projecting them onto another. But the concrete form that each mechanism takes uses one of the *cultural possibilities* available to the individual. Therefore, there are *no* defence mechanisms common to all cultures. The only way to analyse why a concrete attitude is adopted is to use the data of the *relevant culture* to make sense of it.[8]

It has already been noted that every individual plays a multitude of roles. In that case, one problem is that in order for a particular psyche to maintain its integrity, it must be able to find a mutual aspect to all these roles and be able to play them in accordance with it. So how does that work?

In my view, these are the features of this mutual aspect: (a) congruence with the person's logic and interests; (b) being a source of joy to the person; (c) being a source of pleasure; (d) satisfying the person's need for affection; (e) giving the person a feeling of security; (f) satisfying the person's need for approval, protecting them from shame.

Societal roles, simply because they are societal, are performed in a way to satisfy most, or indeed all, of the feeltypes. Therefore, it is important to find out which feature is the active determinant in the choices made by an individual. If an individual has chosen a certain role of their own accord, congruence with their logic or interest is the most important feature. But an individual may also have taken on a certain role based on other features, like being with friends, pursuing a romantic interest, or impressing others. As individuals can satisfy their feeltypes in almost any role, it is *not possible* to leap from the psyche to roles and claim, "A person with this particular psyche would take on that particular role". Individuals may take on a variety of roles, but every person performs each role with the emphases of their own psyche and in accordance with their own subjective characteristics. In other words, being a member of such and such a party, or playing a host of revolutionary or counter revolutionary roles, comes in an abundance of subjective forms.

It is primarily the surrounding environment that determines whether a person takes on a particular role. Their psyche is only a passive determinant. As with other passive determinants, there are exceptional situations in which it can become active; such a case would mean that the person had found the ideal societal role most suited to their psyche. However, this is not the general rule, because it is usually factors dictated by the environment that determine choice of roles, their fulfilment, continuation, or termination.

The individual motor of change is incongruence between the id and superego, which leads to anxiety. In order to cope with this, the ego has to raise the consciousness of both the id and the superego. The more a person is able to achieve this, the better chance they have of becoming an activist able to rescue themselves from anxiety. In doing so, the person sets out along the path to being a social transformer.

6. Influence of Class Position on Individual Formation and Behaviour

The history of modern humankind is estimated to be anything from 40,000 to around 100,000 years long. For much of this huge stretch of time, humanity was organised in the form of small, primitive societies. In this mode of production, known as "primitive communal society", society was basically uniform. It was predominantly collectivist. There was no class differentiation and no division of labour. Societal structures and the individual roles that constituted them were as given, and there was no possibility for people to choose their own roles. As soon as they were born, everyone took on the role ascribed to them by their society. As individuals were cells in the body of society, this obligation would not normally have been experienced as a form of repression. The individual id and superego were mixed together. There was little space for anxiety and contradictions, and therefore no space for individual action and the resulting social change.

As this shows, class is a relatively new phenomenon in the history of the human species, first emerging about 12,000 years ago in the Neolithic age. The foundations for class were laid in the agricultural revolution, when the conditions for regularly producing a *surplus* occurred for the first time in human history. Surplus product is the objective condition for part of a community to be able to survive without working. Class differentiation was built on this foundation, which created the possibility for one class to exploit another. From here onwards, *exploitation* became the most powerful motor of all human development.

Exploitation shatters homogenous cultural patterns based on common interest. It creates conflicting roles and values, severing individuals from compliance with the group and leading to critical questioning. In short, exploitation paves the way to consciousness. This process took thousands of years. The emerging phenomenon of *the state*, over 5000 years ago, laid the cornerstone for the institutionalisation of exploitation.

With the end of the age of barbarism and the start of slavery, or in other words *civilisation*, class division became the most fundamental reality of societies, as it still is today. Ownership of the means of production determines the fundamental nature of not just society's mode of production but also its political and legal structure and dominant ideology. Individuals take their place within institutions, structures, and roles formed inside this general determinative framework. This framework is the most common determinant of individual formation and actions. But the framework is of a *passive* nature. It is an environment, a *framework*. It determines which aspects of individual formation progress with ease or difficulty and which individual practices are executed with ease or difficulty.

The class division of society means that every individual is objectively located in one class or another. Clearly, this will have repercussions. The family structure, the first framework through which individuation-socialisation occurs, and the concrete form of the "child role", are fundamentally based in class, as is education. The professions a child can enter when they grow up are largely connected to class. It is basically class structure that determines the cultural and political roles available to the individual. As already pointed out, this process of determination is usually passive; but it can become extremely powerful depending on the society, the current stage of development, and the social segment occupied by the individual. For example, it can be objectively hard for someone in a conservative village environment to break away from the stereotype patterns that surround them. On the other hand, class may have a lesser effect on determining the individual in the environment of a large city. Individuals belonging to the same class may adopt substantially different attitudes towards culture, art, politics, and many more concrete subjects.

Within the general framework of class, all individuals are the carriers of a particular heredity. They are *actively* determined through *partaking in face-to-face* relations and through the social conditioning that begins as early as the gestation period. This chain of determination equips every individual with a particular character. Throughout an individual's life, the character formed in this way also functions as a *passive* determinant alongside the individual's class.

However, these two passive determinants can become *active* once they reach a certain intensity. The activation of class status in this way relates to the heightening of *class struggle*. When any class becomes aware of their structural position, or in Marxist terminology becomes a "class for itself", class transforms into the most active determinant of its members (or in exceptional circumstances, of the members of another class that identify with the world view of this class).

This transformation is also observed in personalities that are impacted by the class struggle. The six basic feeltypes acquire characteristics that are shaped by class struggle. The individual psyche becomes dominated by affection for others of the same class and hostility towards those in adverse classes. In such a situation, the individual can become an extremely effective player in society, depending on their social position and the number of like-minded people they can organise.

In short, the degree to which class position is determinative depends on social conditions. In normal circumstances, this determination is passive. However, when class conflict heightens, class position can transform into a potent determinant according to the character of one or other individuals.

Finally, it is worth drawing attention to the following methodological reality again: when researching how the individual is affected by basic societal determinants like mode of production or class structure, one of the biggest traps for the Marxist scholar is to jump to conclusions by ignoring unavoidable *mediations*. For example, it is an incorrect methodological approach to say that because the basic reality of society is the mode of production, within which individuals take their place according to their job, the first determinant to be examined in order to understand the individual is the job that each individual is doing;[9] the reason is that this is not how individuals are determined in reality. For it to be possible for an individual to be determined by their job, they would first have to emerge into the world and then later become individuated-socialised. Class position determines an individual through the vehicles of family, education level, culture, and ideology, well before any determinations related to their job. By the time a person takes on their class role in production, their character, or personality, has already largely been determined.

Additionally, as emphasised, both class determinations through unavoidable mediations and class determinations that are manifested more directly within production relations are generally passive. They become active determinants when the class struggle grows; then, the determination is basically effected by passing through ideological and political currents, organisations, social actions, and the media of mass communication (now especially over the Internet). In summary, whether it is activated or remains passive, class determination always arrives at the individual level by passing through a series of mediations.[10]

7. The Unpredictability of Individual Behaviour

Individual behaviour can be examined from two aspects. Looked at from one aspect, behaviour is a manifestation of the individual's personality, which is shaped at an early age by the various influences of heredity, the gestation period, and determinative face-to-face relations that begin from birth. Such factors can, for example, make a person temperamental, assertive, introverted, sanguine, shy, orderly, disorderly, etc. These traits give the manner and tone of an individual's behaviour, explaining how they go about doing things rather than what they actually do. For the purposes of determining the behaviour itself, these traits are passive determinants.

Looked at from another aspect, behaviour reflects the individual's social position. For example, it is a person's class position that basically determines: whether they will go to school and, if so, which school they will go to; which social group they will marry into and at which age; which types of employment they can take; how they spend their free time; which age they can retire; and how they spend their old age. This determination is also basically passive. Class position does not propel a person towards a particular behaviour. It draws the framework of what they will be able to do and how. This framework may facilitate some behaviours and impede others.

Inside this framework, individual concrete actions are generally determined by face-to-face relations with people who hold roles in social structures and people encountered randomly, or by the thoughts, behavioural traits, and fashions dominant in society at that time. These are the main active determinants. For example, voting behaviour is only passively determined by class position. The main active determinants are face-to-face relations and the dominant ideas. Therefore, members of the same class generally only behave in the same way if class consciousness is highly developed among those who hold roles in the social structures that the individual is involved in. As illustrated, it is entirely dependent on levels of consciousness. Otherwise, the individual conforms with the general, dominant ideas or fashions, which are determined by the spokespersons of the dominant class.

The first aspect of behaviour is a matter for psychological study. The object of research is always a concrete individual. Common knowledge of the field can be expanded by analysing how concrete individuals behave in reality. A passive determination framework is established by the tools a child receives from heredity, the characteristics of the body, and the conditions of the surrounding environment. The child's psyche is determined by their face-to-face relations within this restrictive or facilitative passive determination framework. This determinative process occurs on the invariable six basic axes of every psyche. It is the culture of a society that defines how the behaviours on these axes are enacted. For example, the feeling of shame exists in all psyches. However, which situations are considered to be shameful for which people, how the feeling of shame is manifested, which stereotypical behaviours can relieve the distress created by it, etc., are all culture-specific. The individual psyche inhabits the individual body. As the psyche is societal, so must the body be societal. For example, if the feeling of embarrassment in one society is manifested by looking down at the ground, a timid psyche will inhabit a body that is stooped towards the ground.

The other aspect of behaviour – what is actually done rather than how it is done – is a matter for sociological study. The important factor here is the behaviour of role-holding individuals, the masses, groups, leaders of opinion, and those who mobilise social structures. Once more, there are two types of determinants: passive and active. Determinations arising from the individual's social position, roles, the social structures they belong to, their profession, class, skills, and knowledge are passive as long as they remain at a normal intensity. The main determinants affecting the individual within this impeding or facilitative framework arise from other people or thoughts. As other people are normally the other role holders in the social structures that the individual is part of, and they themselves are to some degree affected by their own position, it can generally be expected that people within similar structures have similar behaviours. However, in this scenario, the most influential factor is the general state of society, i.e. the conjuncture. Let us say for instance that there is an epidemic, environmental disaster, war, or temporary crisis – in such cases, the active determinant of the behaviour of the majority will be the circumstance itself. However, in calm periods, the structural positions of individuals, especially the dominant ideas of society, are more effectual.

As the active determinant of societal behaviours is always other people and ideas that have emerged in the past or present, it is *impossible* to predict social behaviour. Nevertheless, it is possible to direct it to a certain degree, for example with advertising. Advertising campaigns are based on research into what things affect individuals and how. The same is true of political behaviours. For example, companies are paid to prepare propaganda on behalf of politicians, with varying degrees of success. This holds also for revolutionary parties. There is such a thing as the right strategy and the right tactics; what defines "right" is the ability to grasp the variables that can determine the political behaviour of the masses.

Sweeping generalisations or predictions about the future of an individual or society, except for those suggesting the general direction of evolution, are not scientific. Nonetheless, a particular prediction may come true if it is based on a sufficiently powerful belief, which then encourages those who made the prediction to claim it is proof of its scientific worth. But regardless of how successful it is, *no definite prediction can be scientific*, because there is no science-based theory that allows such a possibility. As Marx said, however much behaviour may be determined, in the final analysis people make their own history according to their own preferences.

8. Genesis of the Individual/Society: Evolution[11]

Around five to six million years ago, the hominin family diverged from the hominids, whose most important members were the orangutan, gorilla, chimpanzee-bonobo, and hominin. By the time our close ancestors, the *Homo* genus, emerged about two million years ago, many intermediate genera and species had evolved and became extinct. The revolutionary transformation on the road to humanity was bipedalism. Bipedalism had become fully established by about 2.5 million years ago, thereby completely liberating the hands.

The oldest stone tools to have been discovered date back about two and a half million years. They are crudely shaped tools, about the size of a ping-pong ball, obtained by striking pebbles, quartz, and quartzite lumps with another stone. Although extremely primitive, they nevertheless show a revolutionary transformation. "In natural environments, primates are able to use objects around them such as pieces of stone or branches for specific purposes (hunting, defence etc.). ... Chimpanzees have no tool production technology. In other words, they have no concept of design" (p. 51). As Marx said: "...what distinguishes the worst architect from the best of bees is that the architect builds the cell in his mind before he constructs it in wax."[12]

Our first ancestors took a stone lump they envisaged as a tool, struck it from a particular angle using another hard stone as a hammer, skilfully hitting it to produce one or two-sided flakes (p. 91). When the chipping process was finished, the sharp-edged chopper of the habilis was ready to use. ... As food could now be prepared by crushing, breaking, or dismembering, there was less burden on the teeth and jaws. ... "The flakes produced when making tools from the stone lumps were also worked and used as knives. These small cutters were held tightly between the thumb and forefinger. Effective hand-eye coordination and spoken language were essential while making these tools" (p. 92).

The larynx is in the upper part of the nasal cavity in all mammals except humans. Thus, all mammals other than humans are able to breathe while drinking. Human juveniles are the same, having the ability to suck the breast without suffocation. However, the larynx descends around the age of eighteen to twenty-four months, arriving in its adulthood position by the age of fourteen. "The large larynx formed in adults through this process makes talking possible (pp. 97–98). Studies of fossils have led to the conclusion that, ... all species of the human genus had the ability to speak." (p. 99).

"The fourth period, which started almost two million years ago, witnessed momentous climate changes. The ice ages began, affecting a significant proportion of the northern hemisphere. Every glacial period was followed by a rainy and temperate interglacial period" (pp. 106–107).

"About 1.5 million years ago, a new symmetrically conceived, triangular hand-axe entered the human toolbox. This tradition continued until 200,000 years ago. These almond-shape or triangular axes were generally about 5 cm to 35 cm in size" (p. 110). They were particularly useful for hunting, especially for attacks on large herbivorous mammals.

"The Middle Pleistocene era is extremely important to human evolution, because in this time period, which spanned between 780,000 and 127,000 years ago, we witness the evolution of two human species. One was *Neanderthal* and the other was the *sapiens* species that we belong to" (p. 116).

"Evolution into primitive forms of the *Homo sapiens* species began about 200,000 years ago. Pre-*homo sapiens* forms slowly died out and their place was filled by humans that were culturally and anatomically more developed. ... There were always intermediary forms, which also provide the greatest evidence of human evolution" (p. 116). The human species reached its current brain size (1350 cm^3) 150,000 years ago. Since then, the brain has not grown in size proportionally to the body.

Neanderthals and their contemporaries created the stone industry known as Middle Palaeolithic. Mousterian technology is the most well-known phase of this culture. Middle Palaeolithic industry is known for the various types of tools produced.

"A completely new cultural phenomenon occurred along with the *Neanderthals*. This was the concept of the next world. These humans used to bury their dead. A society that buried its dead emerged for the first time in two million years of human history. In the eyes of the *Neanderthals*, death was not the end, it was simply a change of place. Approximately fifty

Neanderthal graves have been found. The oldest of these dates back 95,000 years" (p. 129). Some *Neanderthals* in Western Europe sanctified cave bears, and their skulls have been found inside graves. Based on findings of necklaces made of animal teeth, ornamentation had also begun (pp. 130–131). "Caring for the sick and old enabled *Neanderthals* to live until the age of 50 to 55" (p. 132).

"The *Neanderthals* could not be expected to compete with the culturally more advanced, technologically superior *Homo sapiens* (*Cro-Magnons*)" (p. 133). The Neanderthals were wiped out. The most recent *Neanderthals* date back 33,000 years ago. Özbek asserts that the *Neanderthals* and the *Cro-Magnons* never mixed (p. 137). However, the latest genetic research suggests that we did mix with the Neanderthals, albeit to a small extent.[13] This is of great interest as it means that such commingling between different species is possible, although somewhat exceptional. In fact, there may not have been very distinct differences between species in the evolutionary process. This indicates that there were other intermediary steps on the way to the modern human.

"Modern humans (*Homo sapiens*) … appeared on the European stage about 40,000 years ago. In this age, the northern plains and Alpine region of Europe were covered with glaciers. The climate was as cold as in the time of the *Neanderthals.* The temperature was constantly below freezing point and winter lasted 10 months. In the ice ages, much of the oceanic water was retained in ice masses formed on the continents" (pp. 144–145).

"The transformation from the anatomically archaic human line to a modern human type paralleled the birth of modern behaviour patterns. In particular, the concept of abstract thought and the new behaviour pattern that we know as symbolic expression marked the start of an important turning point in human history. Archaeological findings show that this critical transformation happened long before the migration from Africa. This modern behaviour pattern was reflected in the long term by increasingly significant socio-economic, technological, and demographic developments. Regardless of whether this pattern emerged slowly (between 150,000 and 120,000 years ago) or in a rapid, sudden change (50,000 years ago), it may have been the main driving force in the spread of *H. sapiens* human communities to new worlds beyond Africa. Beads made from the shells of sea molluscs found in Morocco, Ethiopia, the Democratic Republic of the Congo, and South Africa, together with ornaments painted in red or notched harpoons made from bone or horn, are the strongest proof of the symbolic expression and abstract thought that symbolise modern behaviour patterns. They are also considered to be the earliest examples of modern behaviour ever discovered. The latest radiometric dating suggests that the modern thought structure began to blossom 80,000 years ago" (pp. 149–150).

The bow and arrow were first made 20,000 to 17,000 years ago. "Ninety-two types of stone instruments have been identified from throughout the Upper Palaeolithic Age [between 35,000 and 12,000 years ago]. ... Humans of the Upper Palaeolithic age developed tools for working on bone, horn, ivory, skin, wood and soft stones. The sewing needle was made 17,000 years ago. It was between 2–10 cm long and carved from bone or ivory" (p. 151).

Dogs emerged 15,000 years ago in Siberia as a result of the domestication of a type of wolf (p. 152). The *Cro-Magnons* were good enough hunters to have an impact on their environment. "By the end of the Upper Palaeolithic age, about fifty herbivorous animals became extinct" (p. 153). Fishing was also highly developed. For the first time, humans were able to survive until the age of sixty. Fertility periods also grew longer. The population grew. Relations between different communities grew. The human geographical spread widened. Around 30,000 years ago, humans crossed from Korea to Japan and from Siberia to Alaska.

"Humans began to move from cave living 25,000-30,000 years ago, building huts that were designed to cater for all their needs. These huts were big enough to house an average of 20-25 kindred people (4-5 families), with a hearth for each family. ... Like the *Neanderthals*, humans of the Upper Palaeolithic Age buried their dead, also believing in the concept of an afterworld. ... They sometimes left the bones of animals like mammoths and reindeers or ivory figurines in the graves. ... However, they did not make special cemeteries for the dead (necropoles)" (pp. 154–155).

"Together with the *Cro-Magnon* humans, a new phenomenon known as *art* begins to emerge in the Upper Palaeolithic Age 30,000 years ago. ... Upper Palaeolithic humans drew pictures on walls in the darkest, most secluded corners of natural cave corridors" (p. 155). "So far 70,000 of these works of art from prehistory have been discovered" (p. 162). "In this era, there was still no writing. Our ancestors usually expressed their feelings and thoughts through drawing and music. Music, another branch of the fine arts, also emerged in the Upper Palaeolithic Age. Flutes made from the long bones of animals, like deer, and punctured with finger holes have been found in France, Ukraine and Austria. These musical instruments were used by the *Cro-Magnons* 30,000 years ago. In recent years, other similar findings have been discovered, such as flutes in southwest Germany... which date back 35,000 years and are the oldest known instruments in the world. Made of ivory and the radius bones of vultures, the flutes are punctuated at regular intervals with holes. ... In the Upper Palaeolithic settlement at Mezine in Ukraine, the remains of wind, percussion and stringed instruments were found. So it seems our stone age ancestors had an orchestra" (pp. 163–164). "Small portable arts products are seen throughout the Upper Palaeolithic age" (p. 166). "Humans of the Upper Palaeolithic age did not know how to make pottery or earthenware, but they may have been able to fashion dishes from leather, wood or plants. ... Having discovered the needle, our ancestors assembled animal hides and sewed them together like fabric to make clothes of different styles" (p. 168). "They decorated their animal hide clothes with seashells, bone beads or animal teeth. ... Female ornaments also flourished in the Upper Palaeolithic age" (p. 169).

"In Europe ... a deterioration can be observed in cave picture art beginning from 12,000 years ago. The typical animals of the last ice age, such as bison, steppe horse, hairy rhino, mammoth, cave bear and reindeer, which held an important place in the life of *Cro-Magnon* humans, slowly died out and cave picture art was also lost to history" (p. 170).

(This was followed by the New Stone Age/Neolithic, or Agricultural Revolution.)

The following conclusions can be drawn from this summary of the evolutionary line. The first momentous divergence in human evolution came five to six million years ago when the hominin family made a definite split from the hominid families. The main characteristic of the divergent branch is bipedalism. There are numerous reasons why natural selection may have supported this characteristic: being able to see further, better for observing the environment, the ability to carry foraged food in the hands, better potential for effective self-defence, etc. For us, the most revolutionary outcome of the adoption of bipedalism was freeing up the hands. Engels explains the significance of freeing up the hands at great length.[14] Indeed, liberating the hands meant the creation of the "tool of tools", the primary condition of "production" activity, the thing that makes humans human in the broadest sense.

These newly liberated hands only began to make the first stone tools after a 2.5-million-year evolution, i.e. 2.5 million years before our time. But adaptation of the hands was in itself not enough to produce tools. For this process, the brain must also have acquired a "design ability" for the first time. As Marx emphasised in the observation referred to above, the hallmark of humans is their ability to mentally conceive the tools they intend to make with their hands. Probably, the evolving humans, just like chimpanzees, repeatedly tried small changes to adapt the objects around them for their own purposes, progressing with the adaptation in small degrees (transformation of quantity to quality), eventually crossing the threshold of shaping a new object for an intended purpose. Individuals who succeeded in this were mimicked by those around them. Once the technique was established, those in the know taught it to the young and others who were unaware of it. Tools paved the way for humans to meet needs they share with animals, such as eating and self-defence, through social interaction. In other words, with the production of tools, a community of individuals emerged that could only regenerate itself through exchanges with other individuals; thus, society was born. *The individual and society emerged from the same process.*

With the advent of tools, the activities necessary for the survival of individual organisms, such as hunting, breaking open food husks, extracting bone marrow, pulling up roots, or breaking down tubers, were transformed into pre-planned collective activities. In other words, they were transformed into social activities, just like tool making itself. As Marx stated, society came about from the reproduction of material life: "By producing their means of subsistence, men are indirectly producing their actual material life. … As individuals express their life, so they are. What they are, therefore, coincides with their production, both with *what* they produce and with *how* they produce."[15]

The "individual that is a member of a society" is not just a creature with hands predisposed to work and a brain predisposed to design. Through the reproduction of material life, individuals were in an affective/intellectual exchange with each other. This affective/intellectual exchange required a certain psychical development, in the same way that the muscle structure of the hand had to be amenable to the skill of making tools and the cell structure of the brain had to be amenable to the skill of design. Within their social networks, individuals may like (or anger) each other, they share enjoyment (or sadness), they trust (or fear) each other, they approve (or disapprove) of each other, they share pleasure (or pain), they learn together or teach each other things (or share the distress of not comprehending what is going on). As interactions occur along these axes, a behaviour pattern crystallises in each individual. The psyche shapes the body of every individual along with its unique character. The psyche is continually moulding the body through the characteristics that are unique to each individual, for example their stance, gait, poise, call, sitting pose, eating style, embrace, etc. The psyche is the unique configuration of the body, acquired from the network of human relations.

The liberation of the hands had two significant consequences: (1) as much of the work previously done by the teeth was transferred to the hands, the muscle connections surrounding the skull gradually became redundant; (2) the development of hand skills stimulated the growth of the brain in size and complexity. The brain designs, the hands do the work. As the hands work, they acquire new skills in the process. The more advanced tools produced by these new skills need to be designed by the brain. With the production of more advanced tools, the hands learn yet more new skills, etc. Tool production began almost 500,000 years *before* the emergence of the genus *Homo* (a genus including various human species). Oldowan stone work, which was really based on planned design, emerged almost 100,000 years after the arrival of the genus *Homo*.

We know the evolutionary process schema: during the copying of the sequence in the base pair of the DNA helix there are frequent "errors" (replication failures). Some are insignificant and some lead to important mutations. These mutations lead to a particular genetic diversity. The individuals who carry the variations most beneficial to the development of the organism, in other words, most compatible with the environment, are able to reproduce more successfully than the others. Thus, as a result of natural selection, creatures carrying genetic changes more compatible with the environment progress, while others die out. Humans too are creatures that developed according to this evolutionary schema. However, taking into account all the evolutionary events summarised above, the "environment" that determined which genetic variety would prevail in humans was actually society. The survival and advancement of humans as a genus depended on the preservation and advancement of the random mutations supporting collectivity, because the organismic requirements of material life were provided collectively by tool making, collective hunting, etc. Therefore, in the case of humans, the main factor determining natural selection was *sociality*.

9. The Simultaneity and Identicality of Individual and Social Genesis

When it comes to discourses on humans (individuals, society), the simultaneity and identicality of individual and social genesis is the phenomenon that the predominant metaphysical (in the Hegelian sense) thought model struggles most to grasp. Yet, unless this is clearly understood, it is impossible to understand humans. The type of questions always asked are "How did individuals become socialised?" or "How did individuation occur within society?" As long as this is the line of questioning, there is *no escape* from the impasse.

The ideal way to grasp how individuality and sociality emerged as *two faces of the same process* is to look at evolution. Individual-society identicality can be better visualized by drawing on the evolutionary process summarised above.

Chimpanzees, our closest relatives, are also able to use tools when necessary. For example, they can use a stick to bring down a banana, poke something into an insect nest to drive out their prey, or use a stone to break open a walnut. Almost 2.5 million years ago, a group of hominins took tool usage one step further by discovering how to shape stone lumps by striking or scraping them against each other. Who made this discovery? Of course, it was one individual or another. This finding was shared with others by demonstration, imitation, explanation, or learning. The sharing was inevitable, because humans were already creatures that lived in communities. They would not have been able to survive otherwise. So what happened? They began using tools to meet some of the needs they shared with other animal genera, such as eating and self-protection. These tools were initially made through processes that came about accidentally. Then inevitably, once the processes were repeated, the tools were made through *design*. Tools changed both the environment (stones are broken, shaped, etc.) and also the humans that used them (the hand adapts to the tool, the brain learns how to design). The individuals that entered into these relations (the relations necessary for making tools and obtaining the necessary objects by using these tools) were individuals on the first step of the humanisation process. The parties to these relations (individuals) began to form a society among themselves. As they behaved like individuals, inevitably intra-community relations turned into intra-society relations. For example, unlike with bees, these relations are not those that the species is naturally involved in, they are relations that are entered into through practices that are differentiated on an *individual level*. In other words, the individual that is a member of a species becomes a social individual because they begin to behave as *members of a society*.[16]

With the beginning of behaving as an individual, some sensibilities (e.g. curiosity, fear, repulsion), which are imprinted onto the nature of the species, acquire form and content amenable to social relations. Individuals experience curiosity, enjoyment, trust, distress, etc., reciprocally or together. The psyche forms as individuals experience these feelings. Social exchange is echoed in the individual psyche and continually shapes it. This shaping is reflected in an individual's looks, stance, and facial expressions. Psychical shaping at the same time becomes physical shaping. *In short, every individual action constitutes, at the same time, a societal movement, a psychical occurrence, and a physical shaping.*

Meanwhile, evolution continued without pause. The individuals most successful in the collective production process were those whose genetic mutations were best suited to the conditions. They multiplied. Those whose genetic makeup was inadequate gradually decreased or were wiped out by superior species. As shown, social relations pointed out the *direction* of biological evolution, resulting in the emergence of species more suited to socialisation. *H. sapiens* is the last example. However, there is no reason why evolution should stop. On the contrary, it is easy to envisage that the dizzying pace now reached by cultural evolution is impelling biological evolution more than ever.

The historical materialist viewpoint of human genesis can be stated as follows: "Man gained his identity by providing for his needs, which were originally common with those of animals, through ways which led to changes in nature, including his own nature."[17] As a result, one animal species, the hominins, evolved firstly into the *Homo* genus and then, still within this genus, into today's *H. sapiens*. The outcome of this evolution was *not* another new, more developed animal genus; a totally new genus emerged. There is *no other example* of this in nature. As the chief determinant of this was culture, and communication had by now irreversibly come to include all humanity, there was also no possibility of another species appearing within the genus. Humans are a *sui generis* living species: *they are the only creatures whose nature is social.*

One of the most interesting characteristics of this living species is that, by nature, its biological evolution is guided by social needs. This is clear from the general line of evolution. Production of the flint tools that were first produced 2.5 million years ago necessitated the liberation of the hands. (Standing upright first appeared 5–6 million years ago.) Around 2.5 million years ago, the upright stance and bipedalism that freed the hands became the only option. By enabling processes like crushing, breaking, and tearing, tools lightened the strain on the teeth and jaw; the muscles around the skull weakened and the brain began the developmental process (growth in size and complexity) necessitated by tool design. (At the time, the human brain was 660 cm^3, less than half the capacity of the brain today.) Four main tool types were developed in the Oldowan stone work that emerged between 1.9 and 1.4 million years ago: hammer, unifacial chopper, bifacial chopper, and assorted flakes. Examination of these tools has shown that the right hand was used to make and operate them, showing that functions were differentiated among the hemispheres of the brain. The fact that the small flakes were gripped between the thumb and index finger supposes not only visual coordination but also some kind of spoken language used to share this technical progress and pass it from one generation to the next.

The structure of the larynx enables human speech. This ability, which is not shared by any animal species, began to evolve almost two million years ago. As social communication persistently supported evolved laryngeal forms, the physical system on which verbal communication is based attained the level of complexity that allows today's spoken languages almost 100,000 years ago.

Triangular stone axes up to 35 cm long were first made about 1.5 million years ago. This weapon, unequalled in its utility for helping humans to attack and kill large herbivorous animals en masse, was used until 200,000 years ago. The triangular hand-axe, which was used for almost 1.3 million years, is a very efficient tool, especially for hunting large herbivorous animals that are a good source of food. This type of hunting is a completely collective activity. Such hunting operations would be impossible to achieve without sophisticated collective skills, like planning, control, coordination, fast communication, solidarity, surprising an animal en masse, and attacking altogether at the right moment.

After the emergence of these new tools and weapons, there was no compelling social need for the human body to evolve in anyway. (Our bodies took on their present-day appearance about 1.6 million years ago.) By comparison, it took another 1.4 million years for our skulls to reach their current form, which was arrived at in the last 200,000 years.

The focal points of human evolution were in the skull and neck: the brain, pharynx, and larynx. The reason is evident. The motor of social progress was the relations surrounding new technologies. New tools and complex relations required a sufficient level of thought and speech for them to be designed, then used and developed collaboratively.

The discovery of fire at least 500,000 years ago was a momentous driver of humanisation. Fire was good for cooking, keeping warm, providing light, warding off wild animals, making spears from fire-hardened sticks, etc. Cooking allowed the consumption of many objects that had not previously been eaten, creating a whole range of food sources that could now be digested, such as rhinos, elephants, hard tubers from under the earth, and a variety of fish. Moreover, lamps could be fed with melted fat, and hides could be used to make clothes, shelters, and even canoes. It would be impossible for all this to happen without some increasingly sophisticated exchange of thought and knowledge-experience. In these social conditions, evolution ensured the superiority of more intelligent, cunning, and resilient human groups.

When social relations and the thoughts and vocal expressions that reflected them reached a certain complexity, quantity once more turned to quality, leading to the emergence of abstract thought. Burying the dead is an obvious indicator of the development of thoughts that are independent of objects. The oldest graves so far discovered are around 95,000 years old, showing that people at the time had reached the stage where they could conceive of an after world. Almost 40,000 years ago, the faculty for symbolic thought also emerged. People began to draw pictures, make music, and wear ornamentation.

The skill of thinking developed within social exchange,[18] and the emergence of abstract and symbolic thought considerably accelerated tool and production technology. A discovery between about 20,000 and 17,000 years ago, the bow and arrow, dramatically increased the productivity of hunting. Around 15,000 years ago, some wolf species were domesticated, resulting in the emergence of one of humanity's greatest assistants, the dog. Around 30,000 to 25,000 years ago, people began to abandon cave living and construct huts to live in. Thirty thousand years ago, this resilient and skilled human species spread from Indonesia to Australia, from Korea to Japan, and from Siberia to Alaska. Modern humans, who had first emerged in Africa almost 200,000 years before this, gradually spread across the whole world, wiping out all other *Homo* species.

By the end of the Palaeolithic age, 12,000 years ago, the human species had learnt how to design and make ninety-two types of stone tools, which enabled them to process bone, horn, ivory, leather, wood, and soft stones. Mousterian technology, the last stage of Palaeolithic industry, produced knives from flakes and bone, as well as side-scrapers, pointed scrapers, saw-like scrapers, piercing tools, and handled or unhandled triangular points. The sewing needle was first made 17,000 years ago and used to join hides together to make various garments. This technology became so efficient that by the end of the Palaeolithic age (around 12,000 years ago), the human species had already entirely wiped out almost fifty of the gigantic herbivorous animal species they had depicted in wonderful drawings 40,000 to 30,000 years ago.

In order to comprehend the formation of the individual/society within all these processes, the crucial point is to capture the way in which the individual psyche and body developed. A society made up of individuals who make tools and use them to create the objects they need – in other words, a society engaged in *relations of production* – builds a dense, complex feeling-thought network. Latent individual affective/intellectual potential began to surface through the processes of tool making, gathering fruit or tubers, joint hunting, sharing out harvested foods, burying the dead, participating in traditional ceremonies, raising children, etc., and take forms congruent with such activities. In the feeling-thought field, every individual has abilities that belong to six principal axes. We know what these feeltypes are: love-hate, trust-fear, approval-shame, happiness-sadness, pleasure-pain, comprehension-confusion. The network of social communication determines the general behaviour patterns within that society, i.e.: what will be liked and how; which facial expressions, behaviours, and words will be used to express hate or dislike; who can be trusted for what, who will be feared and why; which behaviours will invite praise and which will be subject to ridicule; how joy and sadness will be experienced (e.g. with ecstatic laughter, or mourning with convulsive sobbing, beating the chest or wailing); which things will invoke pleasure (e.g. the drugs or pleasure-inducing herbs used traditionally in many cultures); what will hurt individuals so much they want to die; the ways of exploring the environment, and the means resorted to in order to overcome the confusion of the unknown, for example sorcery. These patterns place an obligation on individuals to conform, which may be very strict or relatively loose, depending on the society's degree of advancement. As a general rule, individuals conform to these patterns. (If they do not conform, it means the culture is heading for change or complete renewal.)

For individuals to display these feelings-thoughts and the requisite types of behaviour, they must doubtlessly possess an amenable structure and some latent potential that can develop accordingly. For example, a creature who has never known fear does not start to feel fear simply because they have entered a particular social relation. To be able to manifest it when the time comes, they must already carry the potentiality of fear. The potentialities that make it possible for humans (individuals acting as members of society) to possess these abilities on six basic feeling-thought axes are actually sensibilities that formed in the evolutionary process (firstly in species lower down the evolutionary ladder, then) in hominids and hominins, well before the emergence of the genus *Homo*. Therefore, what will set them in motion and how the relevant sensitivity will be experienced are predetermined. These are the same sensitivities used by humans. However, in humans they are *recast* by social relations. Eventually, the *stimulus* and outward *manifestation* for the relevant emotion came to be determined by social relations. If this potentiality, whose origins predate humans, had not been present in humans, it would have been *impossible* from the start for any type of culture to emerge.

Humans, and the creatures that preceded them, carry these sensitivities in their genetic makeup. But there is an important distinction between humans and other living creatures. Humans carry the sensitivities as specific *genetic possibilities. When it materialises and in what form* depends on social relations.

These sensitivities are inherited by new generations in both a phylogenetic and ontogenetic framework. In phylogenesis, the basic possible patterns are transferred; in ontogenesis these transferred patterns make certain behaviour models from close relatives more *likely*. Some children are born with a tendency towards fear, some towards excessive boldness. (Their later life may or may not lead them to overcome it.)

As humans are social (cultural) creatures, human heredity is primarily a cultural heredity. This heredity is present in cultural devices. It is present in thousands, millions, billions of *cultural objects* (carriers, witnesses), starting with the modified flora and fauna of nature, and encompassing everything from archaeological sites to museums, buildings to pots and pans, stationery to libraries, exhibitions to live performances, schools to training programmes, religion to soothsaying, elegies to carnivals, fossils to astrology, dancing to protest marches, love stories to orgies, wargames to aid organisations, cooking techniques to swaddling clothes, musical genres to wine varieties, etc. Moreover, it should not be forgotten that feeling, thinking, and behaviour patterns, which are not recorded anywhere, are also cultural objects.

Every new-born human is exposed to this mass of cultural objects. But the creature that is confronted by this mass is *not* a "blank canvas" exposed to all these influences. Each child comes into the world with both phylogenetic and ontogenetic heredity, as well as traces of the numerous influences experienced while still in the mother's womb. One stark reality must be observed: if humans were not born with the potentiality to be human (individuals as members of society), they could never be human in any way. (However much people have tried, they have still failed to raise chimpanzees, our closest relatives, as humans.)

It should also be observed that culture is not just carried in cultural objects. Culture is also carried by *all people* living in a particular time. All individuals, regardless of their age or other characteristics, living concretely at a particular juncture of history, along with their feelings, thoughts, languages, outlooks, facial expressions, behaviour, clothes, jewellery, hairstyles, pedicure customs, etc., *are specific cultural objects.*

The social structures that shape human behaviours, from malleable habits to fixed institutions, are also specific cultural objects. Therefore, culture is not just stored within things but *also in the body and in behaviour*. It is this that ensures that culture never stands still. It is *perpetuated* and *renewed* by the individuals that carry it. Humans keep on rebuilding themselves (and therefore also society) over and over again.

Section B: Literature

1. Historical Materialism and George Herbert Mead

George Herbert Mead (1863–1931) was a North American psychologist and scholar, known particularly as the pioneer of social interactionism and social behaviourism. Despite producing up to one hundred written works, including articles large and small, not one of his books was published during his lifetime. In evaluating his relevance to our subject here, I refer to the book by Charles W. Morris, published in 1962. It is a compilation of Mead's Chicago University lecture notes and provides a valuable introduction to his work.[19]

In my opinion, Mead made two significant contributions. Firstly, he saw, and repeatedly emphasized the fact that "the person" – i.e. the mind, self, and intelligence – is a *derivative* of social relations.[20] Secondly, he realised that inter-individual symbolic communication was the *primary source* in the formation of the individual (individual psyche). In actual fact, these two contributions are different sides of the same coin. The first emphasises the *phenomenon* (that the individual is a product of society), while the second looks at the *mechanism* by which it works.

Animals also communicate with each other. They do so by sending signals that are mutually understood by other animals, such as barking or chirping. As can be seen, meaning is an objective phenomenon. But, according to Mead, as long as it remains at this level, it is insufficient for the organisation of human society, because at this level it is a one-sided transmission. It is not part of a meaning *exchange*. For such an exchange to happen it is not enough for the recipient of the signal to sense only the meaning that they themselves attach to it. They must also understand what meaning it holds for its sender, so that they can put themselves in their place and give a meaningful reaction. If communication is to be shared more widely than among just two individuals, if it is to be part of a network of societal communication, the signals must turn into symbols that clearly show their own meanings. Only when they turn into common symbols can signals become common units of meaning that are understood in the same way by the whole society. If this *reflexive meaning* did not exist, there could be no meaningful exchange among a group. In the human species, this mutual meaning is provided by the common spoken language of the society. It is through this process that humans are created.

By constantly putting themselves in the place of others in social relations, and knowing that others do the same, humans begin to subjectively comprehend the meanings that emerge objectively among themselves. Individual consciousness is not necessary for the existence of meaning. Meaning is initially an *objective* phenomenon, not a "state of mind" or a thought.[21] By being part of the relations that create this phenomenon, the individual acquires a subjectivity due to the capabilities of the central nervous system. This is the source of intellect and self.

Symbolisation enables humans to attach meaning to meaningless objects. In a primitive physical environment, any object only has meaning to a person if they are affected by it. Humans attach meaning to such an object by treating it as a symbol, and of course, this meaning must be mutually shared. It was in this way that humans became the makers of their own environment. If the world of meaning is prioritised, the environment is a product of the subject. It could be called its creation. (As noted below, such a view can lead to *subjective idealism*.)

A person is involved in almost countless numbers of relations with different people and objects. The ability to conduct these relations depends on being able to understand their meanings, which originate from collective life. The possibility to perceive such meanings is provided by the extremely complex network of the central nervous system. As well as the ability to establish relationships between vastly different points, it can also establish relationships between different times. Based on previous experiences, humans can figure out the possible outcomes of something they have not yet done, and then plan out their new action accordingly.

The ability to play with the dimensions of time allows the individual to postpone an action, while comparing and deliberating the potential results of different approaches to it. The individual learns these various potential outcomes from their social experiences. As these experiences accumulate and spread through society via the medium of language, each individual begins to perceive the society's attitude towards particular practices. As a result of constantly putting themselves in the place of others, and putting others in the place of themselves, individuals begin to act as a "generalised other".[22] This is also how knowledge is generalised; for example, after seeing a succession of red objects, "redness" becomes a "generalised knowledge". Behaviour is generalised in the same way; for example, ethics emerge when behaviour adopted by a large majority of people in a specific situation becomes a generalised attitude for individuals.

As may be seen, for Mead, the psyche in all its aspects is a product of *society*. It is the brain and nervous system that enable this product to function on an individual level. However, it does not originate from the system. The ability of the nervous system to establish connections between different spaces and times allows object images, which are initially connected to a specific time, to become devices of the psyche. The psyche either calls up these images with the memory or uses its ability to play with time to visualise the images of objects that may emerge in the future. Thus, having *internalised* knowledge or attitudes generalised through social experiences, the individual weighs up the potential outcomes of various actions learnt from society, designing the future action to be carried out according to this assessment. This is none other than what we refer to as "intelligence".

Mead's great strength was to recognise clearly that individuation, the emergence of the phenomenon of the individual, *would have been impossible* without the mechanism of symbolisation. Symbolisation endows a particular gesture with a meaning commonly accepted in the society.

These meanings start out as a series of regular associations that naturally arise in social relations. Hence, meaning is firstly an objective phenomenon. It is a regular association between a gesture and the following behaviour. In organisms that have reached a certain developmental level, the individual is able to perceive and internalise this meaning association. However, in order for this possibility to arise, the meaning must already have been objectively established. Put another way, the emergence of the individual inevitably requires social relations to have reached a certain level, which in turn necessitates the advent of a common language based on symbolisation.

Another scholar who clearly distinguished the *foundational power* of symbolisation, i.e. culture, was Lev Vygotsky, who I referred to earlier in the book.[23] It is striking that Vygotsky (1896–1934), the Soviet psychologist who only lived for thirty-eight years, and his contemporary George Mead, who died just three years before him despite being born thirty-three years earlier, did not know of each other's existence.[24] Moreover, their theories make an ideal synthesis. Mead explains very clearly the objective formation of meaning relationships, how social roles are established through the crystallisation of symbolic patterns, and how the individual can only establish their own psyche by internalising these. Vygotsky completes this work in two ways: firstly, by arguing that the symbolic tools that make up culture are a product of historical development, i.e. of the concrete conditions in which people live; and secondly by showing how the psyche is transformed through the use of social symbolisation systems. As can be seen, Vygotsky emphasises *historical origin* and the *role of society* in the formation of the psyche. These are essential for a Marxist explanation of the individual. However, the explanation of how this origin takes effect and also how society transforms the psyche, i.e. the idea that can provide the *key* to unravelling society-individual-society interaction, is the *symbolisation* that is at the *core* of both Mead and Vygotsky's theories.

Another factor that Mead emphatically stressed is that the individual is only able to become *self-aware* by internalising how others look at them and how they *appear* to others. The first step towards an individual being able to see themselves is being able to put themselves in the place of others. The ability of any individual to be an individual depends on being able to put themselves in the position of others. Ultimately, saying that an individual is an individual of *that society* is tantamount to saying that this individual has internalised the views of all the individuals of the society. This is made possible by the more or less stereotypical attitudes shared by the individuals of a society towards objects and people. As already mentioned, Mead referred to this common, widespread attitude as the "generalised other". In the process of *internalising* the "generalised other" every individual *personalises it* in some way. This means that part of the "generalised other" changes every time it is internalised. Thus, Mead effectively demonstrates how society *recreates itself* through the formation of the individual.

How can Mead's theories, as summarised above, be viewed from the perspective of historical materialism? Firstly, Mead's observation that the individual is the product of social relations is fully *concordant* with the materialist view of history. Secondly, his explanation of the symbolization mechanism through which this process takes place is a much-needed *contribution* to historical materialism. In fact, historical materialism recognises that the characteristic differentiating human societies from other societies is that they are made up of *self-aware* individuals. As Marx said, what distinguishes the worst architect from the most skilful bee is the architect's ability to firstly mentally conceive of the object to be made.[25] However, the mechanism through which this occurs is not explained. This is where Mead's contribution is of paramount importance: he explains how the symbolisation that occurs objectively in social relations creates a world of meaning which is objectified with common language; and individuality is formed through using this common language. Consequently, he concludes that society can only exist through the presence of *different individuals* who see themselves as members of the society. If everyone was identical, there would be no individuals. Mead's great achievement was to show that *individual difference* originates from the *commonality of meaning* enabled by symbolisation. It is thanks to this commonality that every individual can put themselves in the position of others and establish themselves as a *specific* individual by internalising the views of other individuals, in other words, of society.

Mead's contribution here cannot be underestimated. Nevertheless, his understanding of the individual is too narrow for the Marxist theory of the individual formulated in this book. In fact, in the eyes of Mead, the individual is essentially a *cognitive* creature.[26] Consequently, Mead reduces the psyche to the Comprehension/Confusion dimension. Whereas, as we know, the psyche has five other dimensions as well as this one. The psyche cannot be fully understood without taking into account its other dimensions of Like/Dislike, Trust/Distrust, Approval/Disapproval, Happiness/Sadness, Pleasure/Pain. A psyche reduced to just its cognitive dimension does not exist in reality. At the very most it would be a robot, not a human being.

Be that as it may, there is nothing to prevent Mead's approach being expanded to the other dimensions of the psyche. Mead was inhibited by seeing comprehension as a *purely cognitive* phenomenon. Whereas, as repeatedly emphasized, just as there can be no pure affectivity completely devoid of cognisance, there can also be no pure thought devoid of affectivity. The founding feature of the human psyche is that it is based on an inseparable *fusion* of feeling and thought. At different times, depending on the situation, either the feeling or thought element might be dominant in the mix, but the other element is *always present*. Consequently, just as the affective element is always present in the Comprehension/Confusion feeltype, the thinking element is an inseparable part of the other five feeltypes. That being the case, Mead's mechanism of "establishing common meaning through symbolisation, objectifying the common meaning through common language, and the individual establishing themselves as an individual by using this common meaning to see the self through the eyes of others" can be applied to *all feeltypes*. Feelings, as well as thoughts, cannot be individuated if they are not recognised. Symbols do not just allude to "intellectual knowledge", but also to "emotional knowledge". A person shouting, "Fire!", is not doing so merely in order to pass on news of the event. They are appealing to the emotions of those that hear them – to their feelings of cooperation (or selfishness), fear (or trust in others), disapproval (or need to be liked), sadness (or joy if those in the fire are enemies), and pain (or pleasure depending on the circumstances). Communication can never convey pure thought or pure emotion. Therefore, the cognitive contribution achieved by communication through common meanings also applies to the other five feeltypes. Common symbols also allude to common feelings. Thus, individuals also learn the affective *meanings* of symbols. Through this process, individuals perceive not just what others *know* of them, but also how they *feel* about them. Through this perception, they become an *object in themselves*. In short, the mechanism explained by Mead for only cognitive formation is valid for *all dimensions* of the psyche. That is why it is so invaluable and represents Mead's *greatest contribution* to the historical materialist theory of the individual.

Nevertheless, it is important to note that despite the efforts he made to be a realist and stay within the rules of behaviourist objectivity, Mead sometimes drifted into *idealism*. A notable example is the above claim that "humans are the creators of their own environment". Mead makes the conclusion that the environment itself *is entirely cultural*, based on the fact that the individual can only comprehend their environment through symbolisation. This assumption is the source of his *subjective idealism*. It is true that without symbolisation, i.e. without the common language of society, humans would be *unable to give any meaning* to social relations or objects. But that *does not mean that the natures of all social relations and objects are symbolic*. For example, without the *meaning frameworks* provided by language, humankind would be unable to comprehend or *even recognise* the sun, moon, and other natural objects, neither could they comprehend themselves nor others around them; they would be incapable of giving them any meaning. However, it is wrong to use this fact to claim that if people could not recognise these things, *they could not exist*. This is why, from as far back as the *German Ideology* onwards, Marx repeatedly emphasized that human consciousness does not create the outside world, but rather that human consciousness *originates from the outside world*.

The subjective idealism that at time appears in Mead's work is a common thread in today's *postmodern* approaches. The essence of postmodern thought is to take a characteristic applicable to the individual and *fit it onto society*, concluding that because humans can only comprehend the world through symbols, society is a *cultural phenomenon*. They argue that as culture is created more by language than anything else, society is therefore a "lingual reality". In that case, the only real world is the world of culture.

In contrast, while historical materialism does not reject the idea that the individual's source of meanings is the cultural world, it argues that this world itself is determined by a more fundamental reality – the mode of production. Marxists certainly also use cultural tools (especially language) in support of this argument. But from the viewpoint of historical materialism, the main determinant of social relations is not culture. Culture itself is *determined*. It derives its power from being the *main tool*, the foremost *means of mediation* in the transmission of social reality to individuals.

2. Historical Materialism and Karen Horney

Karen Horney (1885–1952) was the third great scholar to transcend Freud's emphasis on biological impulses and attribute the formation of the psyche to inter-individual relations. Having grasped the value of the path laid down by Sullivan and Fromm, Horney then adopted an independent approach to continuing in the footsteps of these two great pioneers, developing her own distinctive and original theory of the psyche, which was both theoretically productive and highly operative.

While she differed from Freud in both essence and in many details, she saw herself as part of the psychoanalysis current, recognising Freud as its brilliant founder. In her first great work, published in 1937 while Freud was still alive,[27] Horney wrote: "If, ..., one believes that the essentials of psychoanalysis lie in certain basic trends of thought concerning the role of unconscious processes and the ways in which they find expression, and in a form of therapeutic treatment that brings these processes to awareness, then what I present is psychoanalysis" *(ix)*. Her main divergence from the founder of psychoanalysis concerned his belief that these unconscious processes originated from biological impulses. On the contrary, Horney believed that the main determinant of the individual was the inter-individual, and therefore cultural, relations in which they took part.

Like Freud, Horney basically saw the psyche as a system powered by biological energy. The urges that drive the system are unconscious. The individual either does not see them or misunderstands them. For Freud, these impulses originated from the human body, especially sexual drives. But for Horney, this psychical system was powered by anxiety. Anxiety is an unnamed fear of unknown origin, an unfathomable apprehension that is felt deep inside every individual. In Horney's words: "...anxiety is the motor that starts and sustains the process of psychical illness" (1937: 23). Whatever world culture each human juvenile is born into, they suffer a "basic anxiety". This is an inevitable consequence of their objective helplessness as soon as they are born, being entirely dependent on others for survival. The only way to overcome this is to dauntlessly bond with the object of love who enables them to stay alive.[28] People who cannot overcome their basic anxiety in this way will inevitably give into aggressive impulses. Unconscious fear of this strange world creates a hostility towards it, which is in turn unconscious. Hostility feeds fear and fear feeds hostility. If this vicious circle is not stopped, the individual develops neurosis and mental illness. If the spiral gets even worse, the person becomes caught in a cycle of fear and anger that they do not understand, resulting in mental impairment and psychosis.

This basic feeling of helplessness from birth onwards is shared by all cultures, although the patterns of emotional support given to young individuals differ. Some cultures reduce this anxiety, helping the individual to overcome it during the growth process; others actually encourage the anxiety. According to Horney, "Modern culture is based on the principle of individual competition" (1937: 284). This feeds hostility and fear between people, damages their self-esteem, and increases their sense of loneliness. In order to deal with this, individuals engage in an intense search for love-attachment-protection. However, this need conflicts with the fundamental principle of modern culture. As a result, modern humankind is confronted with numerous sources of anxiety, caused by the objective contradictions of the culture as well as the basic anxiety facing every human juvenile right from birth. Those who are unable to cope with this end up in an unconscious spiral of anxiety-anger, inevitably losing their psychical health.

For Freud as well, the basic contradiction between attachment to another and hostility to another is the main factor in the workings of the psyche. Moreover, all psychoanalysts from Freud onwards have believed that attachment to others involves attachment to the self and that hostility to others involves hostility to the self. Freud explains these tendencies with biological drives. The drive to attach to another, "libido", is essentially of a sexual nature, and when directed inwards towards the person themselves, it takes the form of narcissism. The feeling provoking hostility towards others is the death drive ("thanatos"). To explain this, Freud assumes that throughout the life of an organism, there is a constant search for death, an inclination towards lifelessness and escaping from organicity. When directed towards the person themselves, this drive takes the form of masochism. Therefore, for Freud both these conflicting drives are organismic. Yet, there is no organic connection between them: one is not born from the other. Moreover, the assumption of a death drive is not convincing at all. It is no surprise that this hypothesis fell from favour after Freud. Horney, by contrast, asserts that attachment and hostility stem from the same psychical source: anxiety. In this case, there is a real organic connection between them: one is born from the other.

During the first period of her studies in psychoanalysis, rather than explaining any specific type of neurosis, Horney aimed to reveal the character disorders common to nearly all neuroses. The main problem she hoped to analyse, and subsequently resolve, was psychical disorder in adults. While she accepted that such disorders generally originate from a number of negative experiences during childhood, she differed from Freud on two points. According to Freud, the psychical damage experienced in childhood was mainly due to the child's failure to properly overcome the *Oedipus complex*, which pushes the child towards their opposite sex parent. Contrary to this view, Horney sees the *O. complex* as an exceptional phenomenon that may be found in certain cultures. In her opinion, the source of this childhood trauma should be sought, not in the sexual drive, but rather in the child's failure to eradicate basic anxiety. For Freud, this childhood complex takes the form of being stuck in the first three universal stages of childhood sexuality (oral, anal, and genital), i.e. fixation or regression to a previous period. Unless treated, the individual suffers the same symptoms and pain until they die. Yet for Horney, just as there is no such infallible series of stages applicable to the entire human species, also the traumas of childhood cannot be repeated in exactly the same way, because they take new forms that correspond to each person's life. Therefore, contrary to Freud's ideas, childhood experiences are not the most important source for understanding what is happening in the present. These experiences are only worth researching if they really shed a light on today. In her own words, "genetic understanding is useful only as long as it helps functional understanding"(1937: 33).

According to Horney, neurosis is manifested in several symptoms. Firstly, those with neurosis are constantly engulfed by unhappiness and hopelessness (1937: 227). The same is true of those who adopt an artificial joy. Secondly, they are devoid of spontaneity (1937: 115), behaving under unconscious compulsion rather than naturally. Thirdly, their reactions are generally uniform and unrelated to the requirements of the concrete situation. Their reactions, "whether they are appropriate or not", originate from an unconscious defence mechanism. Instead of engaging in an effective mutual exchange in their relationships with others, those with neuroses are motivated by their own unconscious compulsions (1937: 277). When it comes to the behaviours, feelings, or thoughts that create their anxiety, a person with neurosis is always repressed by a series of inhibitions. If these inner obstacles are particularly strong, it can even become impossible to do, feel, or think anything about subjects that are connected with the anxiety. Nevertheless, the person attempts to do so, experiencing a disproportional tension, fatigue, or exhaustion. Activities that a healthy person would usually find enjoyable become the source of distressing anxiety for someone with neurosis (1937: 53–59).

Horney meticulously researched the various ways that people search for solutions to overcome their anxiety. In doing so, she not only contributed to theory but also illuminated the way for therapists and those attempting self-analysis. Horney outlines four main ways in which people in our culture deal with anxiety: rationalising the anxiety, denying its existence, benumbing it, and avoiding the feeling, impulse, or situation that causes the anxiety (1937: 47–48). Examples of the strategies described by Horney are easily found. Let us take the example of someone who is terrified of leaving the house: they can rationalise their fear by using the fact that traffic is dangerous; or they can deny their fear by saying they simply do not want to go out; or they can take drugs or alcohol in order to go out; or they can stay home, never leave the house, nor even think of doing so.

It was stated above that feelings of hostility and aggression originate from anxiety. If these feelings pose a considerable threat to the inner balance, the individual, entirely unconsciously, reacts in a way that suppresses or projects the feeling onto others before even attempting any of these strategies. The advantage of projection is that it provides the person with a means of rationalisation. For example, when someone with feelings of hostility attributes this enmity to others, in the belief that the characteristic belongs to them not themselves, it provides a way for their own hostile behaviour, which they have been unable to suppress or overcome, to be rationalised and justified (1937: 69–70). Projection is of great use to the ill person. When someone projects the feelings or inclinations they are ashamed of onto others, it provides an opening for that person to be harshly critical of such behaviour without actually hurting themselves (1937: 186).

According to Horney, our culture offers the individual four main inter-individual relationship templates for protecting against basic anxiety: affection, submission, power, and withdrawal. They might be summarised respectively as follows: "If you love me you will not hurt me"; "If I give in, I shall not be hurt"; "If I have power, no one can hurt me"; "If I withdraw, nothing can hurt me" (1937: 96–99). In order to escape the trauma of this immense fundamental anxiety, individuals usually adopt more than one of these behavioural patterns. Without question, this leads to unresolved inner contradictions, deepening the neurosis.

The need to be loved and at the same time be powerful against others is both unavoidable and also the main catalyst of inner contradictions. A growing need for affection is an unfailing indicator of the individual's search for support when threatened by personal crisis (1937: 151). The individual tries every strategy in order to obtain affection: offering some kind of profit to another (bribery method); playing for sympathy; using a supposed injustice they have suffered to claim that love is their "right" (appealing to justice method); making various threats (e.g. "I'll kill myself") (1937: 139). If necessary, they also resort to self-blame. This pre-empts any criticism and invites others to console them (1937: 242). The person suffering anxiety is also unable to refrain from seeking power. Power is one of the most effective ways of preventing our wounds being exacerbated by others. Using power also enables release of the hostile feelings that are inseparable from anxiety (1937: 166).

The conclusions of Horney's early research gave her a guide for future study, while also providing others with important clarity both in theory and practice. She arrived at the following generalisation: in reaction to anxiety, people adopt one of three main responses. They either go *towards* others, trying to get closer to them and make a connection, or they move *against* people, trying to put pressure on them, or they move *away* from people, choosing to avoid relationships. Horney's great achievement is to collate numerous attitudes and behaviours, such a complex collection of data, under just three headings.

After this first period came to a close, Horney continued her research, finally presenting her partially refreshed and further developed ideas in a comprehensive new work written two years before her death.[29] The main shift in her views was as follows. In the first period of her work, she saw the chief source of psychical health as inter-individual relations and consequently the surrounding culture; in the second period, she lowered these relations to secondary importance and began to seek the main source of anxiety in a second personality that dominates unhealthy psyches. This second personality is an idealised personality that the individual attributes to themselves in order to cope with anxiety, trying to adapt every behaviour, feeling, and thought to it. This idealised, ill personality tries to strangle the real personality – in the words of Horney, "like a Frankenstein". The idealised personality uses inter-individual relations as its own vehicle. Eventually, the neurosis takes the form of a personality disorder that inhabits the person's relationship with both themselves and with others.

According to her new understanding, at the centre of the psychical system sits a vital conflict between the healthy efforts of the real personality to develop its own potential and the unhealthy efforts of the idealised personality. The destiny of every individual is determined by this conflict. The illness (neurosis or psychosis) is a result of the idealised personality overwhelming the real personality. The only way back to a healthy psyche is to strengthen the real personality enough to overcome the idealised personality (1991: 367–368).

The *idealised personality*[30] is built on a pathogenic conceit. In order to overcome the anxiety created by a subconscious fear of humiliation and abjection, the individual, again unconsciously, establishes an idealised personality that will be undeterred by these feelings. This idealised personality has to conform to numerous strict rules. (Horney refers to these as "inner dictates" or "shoulds".) The individual *should* be able to do, feel, think, know such and such a thing. At the same time, there are numerous other things that certainly must *not* be done, felt, or thought. These obligations and prohibitions are not the product of healthy ethics, but of a pathogenic conceit. As the health of the psyche is damaged, the idealised personality becomes the measure of everything (1991: 110). The act of behaving like this idealised person becomes the only motor moving and directing the individual in their actions (1991: 84). The person loses their freedom of choice (1991: 169).

It is undoubtedly impossible to attain the perfection aimed at by the idealised personality. Outside conditions do not allow the perfection to happen, and the idealised personality cannot completely supplant the real personality. As a result, the *phenomenological personalities* ("actual or empirical selves") that we see in concrete can never quite meet the expectations of this debilitating conceit. The idealised personality reacts to this in various ways, seeing it as a right that the demands feeding this self-glorification are met and trying to impose them everywhere. In the meantime, the individual pressurises themselves even more. When they again fail to achieve perfection, they begin to feel *self-hate*. They humiliate themselves, applying merciless criticism as they torture themselves with blame. At the end of this process, the individual may suffer collapse (1991: 117). Clearly it is impossible to endure such a torturous process, so eventually the ill person blames others for what they have been through, beginning to subconsciously see themselves as a victim. They are surrounded by enemies (1991: 136). Their pain deepens further, but it is not the type of pain that can illuminate the consciousness, it is a blind pain (1991: 163).

One of the foremost needs of humankind is to protect our personal integrity and prevent it from collapse. This need is even more trenchant in an unhealthy person who is wrought by contradictions (1991: 240). The idealised personality tries unconsciously to meet this need for integrity by focussing on one of three main strategies: "expansive solution", "self-effacing solution", or "resignation" (1991: 191). Here we can immediately recognise the threefold grouping that Horney proposed at the end of her first period of research. However, this time the emphasis is on the individual internal structure, rather than the problems caused by inter-individual relations; Horney's priority is to explain neurotic personalities as ill people dominated by what she calls idealised selves, and she achieves it by grouping these selves into a number of main types.

In the works of her later period, as well as acknowledging that different idealised personalities can coexist in real people in a discordant union, she also tried to describe these main types in depth. Using her experiences in psychiatry, she details how the harsh demands, bullying behavioural norms, justifications, and vulnerabilities of each type are exhibited in inter-individual relations. As she unravels the hidden workings of these internal structures, she never loses sight of the fact that every internal factor will be manifested in inter-individual relations (1991: 291). As a result, she makes numerous valuable discoveries that are ideal for fully developing and enriching both psychological theory and psychoanalytical techniques. For example, she presents the three main unhealthy formulas adopted to avoid reality in our current age: "drowning in fun" of some kind, seeking out opportunities to "focus on success", and living "like a robot" without thinking about anything (1991: 286–287).

According to Horney, rebuilding the patterns of inter-individual relations through changes to the social order is not sufficient to remove these unhealthy idealised personalities. There should be a return to the therapeutic wisdom of ancient times and, above all else, inner healing power should be supported for both the body and mind. The approach for mental disorders is for the person to focus on their *real personality*. The real personality is made up of spontaneous feelings that possess both depth and sincerity, while the ill person is repeatedly embroiled in their idealised personality. Most of the time, they seek psychiatric help to overcome the obstacles inhibiting this personality, rather than to free themselves from it. However, the role of the therapist is to try and clear the way for the person to strengthen their real personality, peel off the mask from the idealised personality, and thereby facilitate its removal. In the words of Horney, "It is at bottom this question: does the patient want to keep whatever is left of the grandeur and glamor of his illusions, his claims, and his false pride or can he accept himself as a human being with all the general limitations this implies, and with his special difficulties but also with the possibility of his growth? There is, I gather, no more fundamental crossroad situation in our life than this one" (1991: 356–357).

How can the ideas of Horney, as summarised here, be evaluated from the perspective of historical materialism? On the one hand, her definition of the individual through their social relations rather than biology is fully in agreement with one of the main tenets of Marxism. The emphasis on the internal structure of the individual in her second period of work should not be understood as a deviation from this basic direction, but rather as an attempt to explain the individual arm of the individual-society dualism within its own specificity. Evaluated in this light, her work is an extremely rich contribution to the second level of the three levels of dialectical identity: the level that anticipates research of the specificities of the parties. But Horney went even further, also making a valuable contribution to the third level of "transformation of the parties into each other". She produced seminal studies on numerous subjects, such as how cultural pressures are reflected in the individual character, what occurs in the individual psyche as a result of problems arising from inter-individual relations, and how personalities formed in this way in turn affect inter-individual relations. In other words, she gave valuable clues on how to resolve the problem of "mediations" that emerges at the third level of dialectical identity.

However, the understanding of "human nature" on which Horney based these analyses is incompatible with a dialectical materialist conception of history. According to Horney, there are three fundamental views about human nature: (1) people are bad and therefore must always keep themselves under control; (2) people are a mixture of good and bad and therefore they must rein in the badness and use their will to develop the goodness; (3) people are neither good nor bad, and therefore "goodness" in the sense of human beings means the free development of people's own inner potential (1991: 14–15).

Horney identifies with the third view. Humans develop in accordance with their own potential if they are uninhibited, just like an acorn that falls to the ground and, if left unhindered by adverse conditions, grows into an oak tree. We do not need to explain to an oak tree how it should develop, nor are we able. The individual is the same. Every individual will find their own aims and values in life, depending on: the clarity and depth of their feelings, thoughts, wishes, and interests; the strength of their will; their ability to use their own talents; any special abilities they may have; and their ability to express themselves and to relate to others with their own instinctive feelings. This is how their personality will be established. This ability is common to the entire human species, but the way in which it materialises is specific to each individual (1991: 17).

As can be seen, Horney's understanding of human nature is actually *essentialist*. It is this characteristic of her work that conflicts with Marxism. Such a view of human nature ignores the historicity of society and presumes a common human essence that is manifested separately in every individual. It is true that every person has a different temperament and that they have different capabilities connected to it. However, the forms of materialization that these capabilities can take are, in every historical period, provided by the society of the time. The options of anyone living in a specific society are limited by this framework of possibilities. The ways in which this materialises are different in primitive communal societies, tribal orders, slave societies, feudalism and capitalism, and even for instance between the contemporary USA and China albeit to a limited extent. No individual can emerge with their own exclusive values completely independent of all this. The values that individuals are able to identify with are proffered by society. Any approach to the process of humanisation that starts out with the individual is fundamentally wrong. Despite her valuable contribution to the materialist conception of history, unfortunately Karen Horney too was misguided on this point.

3. Historical Materialism and Harry Stack Sullivan

Although relatively unknown, the North American psychiatrist, Harry Stack Sullivan (1892–1949), is one of the most important names in the world history of psychology. Only one of his books was published during his lifetime: *Conceptions of Modern Psychiatry* (1940). His three most important works, posthumously compiled and published, were *The Interpersonal Theory of Psychiatry* (1953), *Schizophrenia as a Human Process* (1962), and *The Fusion of Psychiatry and Social Science* (1964). Along with Erich Fromm and Karen Horney, he opposed Freud's drive theory, arguing that the main determinant of the individual was social relations, i.e. culture. He is the founder and most important figure in the movement known as interpersonal psychiatry or interpersonal psychoanalysis, which argues that the main determinant of the individual is their face-to-face relations with others.[31] Here, I will only examine the theories of this great thinker in so far as they relate to historical materialism and the feeltype theory.

According to Sullivan, radical behavioural changes occur in many animal species, especially in advanced mammals, in response to radical changes in the environmental conditions. But the species of the animal does not change as a result. Yet, humans are animals that changed species. They first emerged as animals, then through living in a cultural environment they ceased to be animals and became humans. It is fruitless to search retrospectively for the animal at the emergence stage. There should be no quest to attach social dimensions to an animal species. That animal has *vanished*. Its place has been taken by a person, i.e. a human being.[32] While a person also has a biological aspect, that is not what defines them as humans. A person only forms, develops, and survives within a network of relations with other persons. As soon as they are deprived of this network, they immediately begin to lose their human qualities.

A human enters the world of human relations as soon as they are born. Therefore, we should not be misled by Sullivan's statement that "humans are born as a type of animal". Although this statement does not appear to do so, Sullivan's ideas support the main argument put forward in this book. Humans are not an animal species; they are a completely different being to plants and animals. Numerous animal species live communally, therefore it is inadequate to say that this being is one that lives in a community. Indeed, humans cannot develop inside *any* community, it must be within a cultural environment that ascribes meaning to behaviours and objects. It is for this reason that common language, made up of signs with meanings, is the most important vehicle in the process of humanisation. Without language, there could be no humans.

Biological development is undoubtedly a condition of existence for the development of humankind, but it is not its determinant. On the contrary, in some circumstances social relations can be seen to directly affect biological development. For example, it is common knowledge that an extremely problematic interpersonal relationship environment can delay adolescence.[33] According to Sullivan, the effect of social relations on determining the individual can be split into a six-stage process, beginning with birth and continuing until around the age of eighteen. The first three stages altogether last ten years: "infancy", which lasts until the age of two and during which speaking skills are acquired; "childhood", which begins with the development of speaking ability and ends when school begins at the age of five; the "juvenile era", opening with entry into the school environment and closing in preadolescence at the age of ten. The final three stages last eight years: "preadolescence", from age ten until about thirteen; "early adolescence", from age thirteen to fifteen and "late adolescence", from age fifteen to around eighteen.

At all these stages, the determinant is *social relations*. The passage from the first to the second stage comes about with the acquisition of language, the most important vehicle of humanisation. Passage from the second to the third stage is determined by the beginning of schooling, which brings children across a major threshold in the humanisation process. Once the child is at school, their environment is no longer limited to immediate family, relatives, and play friends. They begin to compare their own family with teachers and friends' families. While they are a privileged object of interest at home, at school they find themselves among equals. It is here that they learn the basic rules of societal life. At school, children learn to give and take, compromise, cooperate, and compete. If the child has grown up in a healthy environment, by the age of ten they reach a level of maturity that allows them to feel *secure* in the society. At this same age, children acquire a disposition that manifests itself in all their personal relations and gives them an "orientation in living".[34] After this, the time comes for one of the most fundamental needs of the human species, intimacy. The only way to achieve this is to find a friend, comrade, or kindred spirit of the same sex. At this first stage of youth, the developing human experiences *loneliness*, the most terrifying emotion. The healthy way to overcome this is to establish a close friendship with another person, taking on their needs and making them your own. In adolescence, the need to avoid loneliness becomes more dominant. However, adolescence combined with the associated arrival of lust strains human relations. Lust pushes young people towards those of the opposite sex. But it is generally difficult to reconcile this urge with the need to establish close, secure relationships. Within the network of interpersonal relations, it is only at about the age of eighteen that young people meet the search for the three basic needs of all people – security, intimacy, and pleasure – through their own unique pattern of relations, their *personality*, and thereby pass from adolescence to maturity.

Sullivan believes that a baby goes through three main stages in the process of perceiving their environment. For Piaget, whose ideas were discussed earlier, the process is the same, but it is divided into six stages. The first stage is a wholesale perception. The child cannot distinguish themselves from the world of objects. The distinction between self/environment has not yet formed in their mind. As it may be recalled, Piaget described this situation as the child existing in some kind of "magma". Sullivan called this form of perception "prototaxic". Afterwards, the child enters a stage in which they can distinguish themselves from objects, acquiring the basic dimensions of comprehending objects. In Piaget's words, "During this period, the child has acquired the fundamental dimensions of the relationships of objects to the world: object permanence (the fact that the object stays the same even if it is moved or has been out of sight for some time), space as the context of the object, time consecutive to the movements of the object, and the perception of causality leading to the movement of the object. However, these are still raw. For example, the child initially understands causality with a magical-phenomenological quality. This is because it always sees its own action (the magical aspect) and the proximity of one object to another (phenomenological aspect) as the source of causality."[35] Sullivan made similar observations, stating that children connect their experiences to the results in a magical way, and that while still at this stage, they learn to remember and foresee, to comprehend "before" and "after", and to perceive what has, and what will, happen. But this perception is not rational, it is impressionistic. Sullivan calls this "parataxic" perception. The connections the child makes are still random. However, the child does not stop there. Towards the end of their second year, through trial and repetition, they begin to understand that the phenomena they encounter are in fact the signs or symbols of something else and that *they carry meaning*. Sullivan refers to this as "syntaxic" perception. The most effective vehicle for this understanding is the society's mutually spoken language. By presenting meanings through common signs shared by all, spoken language facilitates the unity of meaning that enables a general understanding in the society. By learning language, children join in this meaning alliance. They begin to explain cause and effect relations through relating them to the common reasoning that is the basis of societal comprehension.

Despite all these parallels, these two great scholars have a different emphasis, because they are really tackling separate subjects. Piaget closely examines intellectual development in humans, aiming to reveal how the comprehension-knowledge ability is formed. Sullivan, on the other hand, is interested in how the main personality characteristics originating from interpersonal relations are formed, how mental illnesses emerge, and how they can be treated. Piaget's main conclusions were summarised earlier. He succeeded in unravelling the developmental stages of juvenile intelligence. His work relates to the Comprehension/Confusion dimension of the feeltype theory. Sullivan, however, was concerned with the other five dimensions of the feeltype theory, particularly the Trust/Fear dimension. For Sullivan, the most important aspect of interpersonal relations is the feeling of trust. When people cannot find this, they become melancholic. A baby is particularly vulnerable to the absence of trust at the prototaxic stage, when they cannot distinguish themselves from their environment and understand everything as a mass of which they are a part.[36] Due to this prototaxic perception, whenever a child affected by such a relationship pattern experiences the stress of insecurity in their later life, they will experience an anxiety of ***unknown origin*** that they "cannot name". If this tension is too traumatic, the individual begins to block out the situations that cause this stress. In the words of Sullivan, they develop a "selective inattention" towards these situations. However, if this becomes prevalent, it leads to disconnections in their personality. This is the cause of many mental illnesses, especially schizophrenia. Sullivan's originality stems from his addition of the "not-me" concept to the commonly known distinction of "good-me" and "bad-me". According to the "not-me" concept, individuals overcome by distress become unaware of some of their words and behaviours. Freud explains this escape from reality as a "suppressing" reaction of the subconscious, which wants to protect itself from the judgement of the superego. According to Sullivan, however, this issue is connected to a kind of alienation. The person becomes alienated from their own words and behaviour, even to the point of developing split personalities.

Sullivan's great achievement was to identify interpersonal relations as the source of all developments in the individual psyche. This is why his work is essentially consistent with the approach of historical materialism, which identifies the chief (or active) determinant as social relations. His views might be summarised as, "what makes me who I am is the attitudes of others towards me".[37] In the prototaxic period, when the baby still cannot perceive of themselves as a separate being, they develop behaviour patterns based on the mother's (or person fulfilling the mother's role) behaviour either as a "trustworthy person" or the opposite, "frightening person". They learn to connect these behaviours with facial expressions and tones of voice. When they enter the parataxic perception period and learn how to distinguish objects, they begin to see their own body as an object. This is the first step of the notion of self. Whenever the mother approaches this body with reassuring, comforting, and affectionate behaviour, the child senses themselves as "good-me". When the mother frowns or raises her voice, the child sees themselves as "bad-me". As long as it remains at low levels, the feeling of "bad-me" has some benefits. In order to escape from the tendencies that provoke those around to make them feel like "bad-me", the child tries to obtain security by increasing its "good-me" tendencies. This is how a child is *socialised*. However, if the "bad-me" perception reaches unbearable levels, rather than continuing to experience or cope with the pain, the child identifies with "bad-me", or begins to develop "not-me" attitudes. Both these reactions can lead to mental illness or even deformities.

According to Sullivan, the main characteristic of the human species is sociality. This sociality is the *origin* of every individual's psyche. The need for inter-personal relations is the most fundamental human need. The human juvenile comes to the world so frail that even meeting their most biological needs necessitates the care of another. In fact, establishing a secure relationship with another comes even before biological needs. Sullivan believed that sociality was not just a characteristic of humanity, but rather the characteristic that makes humans human. It is this characteristic that *creates us* as individuals.

The "feeltype" theory advocated in this book is completely consistent with Sullivan's views. As it may be recalled, according to the feeltype theory, all human psyches have six different feeling-thought dimensions, generally intertwined with one another: Like/Dislike, Trust/Fear, Approval/Shame, Joy/Sadness, Pleasure/Pain, Comprehension/Confusion. Each one of these dimensions is really a feeling-thought *ability*. These abilities are *concretely* defined by the *culture* of the individual's society, and this cultural determination is always achieved through *face-to-face* relations. Undoubtedly, there is also a very comprehensive and important determinative process outside of cultural face-to-face relations. However, the human juvenile *cannot be humanised* without sharing in the human warmth of face-to-face relations. Sullivan makes the same point. According to his viewpoint, the most determinative of these dimensions is the Trust/Fear dimension. This is due to the incredible helplessness of the human juvenile. Similarly, being liked and approved of are necessary in order for feelings of security to flourish. Without doubt, such a relationship would bring joy and pleasure to the individual. In addition, Sullivan emphasised that the ability of "comprehension", which Piaget examined in such detail, was greatly affected by the individual's feeling of security. He was correct. Without security, the "not-me" aspects of the personality begin to dominate. In other words, the individual's relationship with reality is damaged. It is obvious that the learning-understanding-knowing feeltype would be unable to flourish in such circumstances.

In all these respects, Sullivan's ideas are fully compatible with the theory of the individual put forward in this book. Consequently, Sullivan's work is a significant contribution to the development of a Marxist theory of the individual. However, Sullivan was not a Marxist and the difference is this: the determinism of the mode of production, which is argued by the historical materialist viewpoint, is missing from his analysis. Therefore, he does not have an opinion of how culture is impacted by the level of development of the class struggle and the forces of production. To understand how these factors determine face-to-face relations, the proponents of historical materialism must seek out different sources. Nevertheless, Sullivan made an outstanding contribution to the establishment of the affective bridge linking face-to-face relations with the formation of the individual.

4. Historical Materialism and John Bowlby

It was John Bowlby, the British psychologist, psychoanalyst, and psychiatrist (1907–1990), who more than anyone else successfully theorised the basic need for security identified by Sullivan. As stated by Jeremy Holmes in his fascinating book about Bowlby, "Apart from Freud and Jung, Bowlby is one of the few psychoanalysts who have become household names and whose ideas have entered the vernacular. The ill-effects (or otherwise) of maternal deprivation; the importance of bonding between parents and children; the need for a secure base and to feel attached; the realisation that grief has a course to run and can be divided into stages – these are concepts with which people far removed from the world of psychology and psychotherapy are familiar. All may be traced, in whole or in part, to the work of John Bowlby."[38]

Bowlby's interest in the subject was mainly aroused by the mental trauma suffered by children separated from their parents and raised in care institutions, especially during World War II. Although he had been trained as a psychoanalyst, Bowlby saw that the theories of Freud and mainstream psychoanalysis did not have a satisfactory explanation for this phenomenon and sought theoretical support in other fields. His main research areas were evolutionary theory, ethology,[39] comparative psychology, control theory, and information processing. Bowlby's theoretical conclusions are summarised below.

Many animals, not to mention humans, react with fear to sudden movement or an increase in sound, light, etc. Both humans and numerous types of animals give the same reaction of fear when separated from a habitual protective individual (p. 30).[40] The function of this evolutionary reaction pattern is to keep the species, especially juveniles, as far from the possibility of attack as possible. In humans, this protection is generally provided by a mother and father. This foundation of trust is not just to protect the juvenile, it is also necessary for its entire development. A child can only "explore" the outside world if they have this kind of attachment assurance, a haven they can flee to when in danger. Through making journeys of discovery with continually expanding boundaries, the child enables its own development as a member of society (p. 11).

The need felt for the mother-father (or others fulfilling this role) is of such fundamental importance to enabling individual development that it should be seen on a par with eating and copulation for the survival of the species (p. 27). Traditional psychoanalysis made the huge mistake of seeing this healthy and crucial need for attachment as a childish weakness that must be overcome as soon as possible (p. 12). On the contrary, far from eradicating it, such a fundamental foundation of trust is an essential condition for everyone young and old to develop a healthy, stable personality throughout their life (p. 46).

The individual's relationships with others primarily take their shape from the relationship model formed within the framework of this fundamental trust relation. There are three main patterns encountered in the attachment relations that someone can form with the one or two people closest to them: (1) secure attachment; (2) anxious resistant attachment; and (3) anxious avoidant attachment. The first pattern is healthy. It enables the individual to feel self-assured and, based on this assurance, to establish relationships that will enable them to continually develop. The second and third patterns are unhealthy. In the second pattern, attachment takes the form of a continually fractious relationship. In the third, the only way to escape the anxiety created by the attachment pattern is to avoid intimate relations. As well as causing a continual dissatisfaction and stress in the individual's personal life, these unhealthy attachment models can also lead to social relations marked by aggression or indifference (p. 123). In addition to these unhealthy personality characteristics, if the fundamental trust relationship is chronically lacking, the individual can also suffer from severe mental illness.

People cannot consciously control the reaction and evaluation mechanism they develop towards their closest relations within their experience of life. The "unconsciousness characteristic" argued by psychoanalytical theory is also applicable here. We avoid stimuli that cause anxiety, or if necessary develop unconscious defence mechanisms against them. When confronted by certain stimuli, the internalised relationship pattern that we develop due to relationships with those closest to us, what Bowlby called our "behavioural system", either enters a relationship or withdraws depending on the circumstances. Withdrawal is unconscious. According to Bowlby, this type of unconscious withdrawal can lead to severe mental illness, just like the concept of "selective inattention" mentioned above, which played an important role in Sullivan's schizophrenia research.

Bowlby's explanation of aggression, developed in the context of attachment need, is particularly noteworthy. As known, Freud's explanations for aggression, such as the death drive, were particularly unconvincing. The argument that aggression was instinctive was later defended in the framework of ethology studies. According to this view, it is a functional behaviour developed through evolution. Another view, which also originated from Freud, argues that aggression is a reaction developed by the individual in response to obstacles. Yet another approach sees aggression as a behavioural pattern learnt during the social learning process. On the other hand, for Bowlby, an important source of aggression is the individual's reaction to the potential disintegration of their attachment framework. This explanation is particularly interesting in that it sheds light on the reasons for aggression towards those who are closest. According to Bowlby, the attachment need is of such fundamental importance to the individual that it can cause a huge surge of anger towards the person who is the main focus of attachment if there is a threat that the person may leave or abandon the other (p. 30). This same anger may lead to suicide in someone who is unable to endure the pain of separation.[41]

The fundamental attachment pattern shapes the individual's entire life, and more importantly also affects the next generation. Over time, a child who is unloved, rejected, and maltreated by their parents (or those fulfilling the parental role) develops an unhealthy personality. They either acquire a negative, confrontational behaviour pattern or withdraw from those around them as much as possible. When people conditioned in this way become parents themselves, they have a tendency to treat their own children the way they were treated during childhood. Consequently, deprivation of the fundamental need for trust is generally passed down through generations, leading to personality disorders (p. 37).

Bowlby's theory is far richer and more elaborate than the rough summary presented here.[42] The attachment/separation relationship pattern is not an explanatory model with just one variable. A realistic explanation should take into account not just the child's personality, but also the personality of the person who is the subject of the attachment/separation, together with concrete events and the way they are perceived, and the social environment.[43] Seen in this light, maternal deprivation, albeit of crucial importance, is just one of many factors in a complex web of influences leading to vulnerability, rather than the direct cause of neurosis.[44]

What is the value of John Bowlby's attachment theory to the materialist view of history? For one thing, by seeing personality as primarily a product of the environment rather than of instinct or heredity, his ideas support the argument that humans are a social species. Moreover, Bowlby succeeded in building a comprehensive empirical and theoretical scientific framework for this viewpoint, which was also strongly supported by Sullivan. For Bowlby, sociality was the foremost factor in the formation of the individual and their behaviour patterns. Within this sociality characteristic, Bowlby focussed on studying the emotional bonds between the individual and those closest to them, their formation, development, and severance. Even more importantly, he constructed a theory around them, thereby analysing the core relationship of sociality in interpersonal relations. This is the magnitude of his contribution. Bowlby made no attempt to construct a general theory of human sociality. However, he did explain the "attachment pattern", the most crucial aspect of sociality with regard to the individual psyche. He did this with a theoretical structure that was constantly elaborated with observations and experiments. For all these reasons, John Bowlby made an immense contribution towards a historical materialist theory of the individual.

5. Historical Materialism and Erich Fromm

Erich Fromm (1900–1980) had a closer connection to Marxism than any other scholar rooted in psychoanalysis. For this reason, he is extraordinarily important in a study that attempts to construct society-individual identicality while preserving the specificities of both poles. Until now, the most satisfactory theoretical explanation of the society pole of society-individual identicality has been provided by Marx, while Freud provided the theory of the individual pole. Fromm was very familiar with both these thinkers and tried to fuse their ideas in an original synthesis. Therefore, his work is a source that must be carefully examined here. Furthermore, through his efforts to create this synthesis, Fromm acknowledged that Marx was a more far-reaching thinker than Freud when it came to the society-individual phenomenon, seeing historical materialism as the chief determinant in its relationship to psychoanalysis. In *Beyond The Chains of Illusion: My Encounter with Marx and Freud*, published in 1962, in which he attempted to fuse the ideas of the two great thinkers to whom he devoted his life, Fromm wrote: "Marx had a much deeper insight into the nature of the social process, and he was much more independent than

Freud of the social and political ideologies of his time. Freud had a deeper insight into the nature of the process of human thought, affects, and passions, even though he did not transcend the principles of bourgeois society. They both have given us the intellectual tools to break through the sham of rationalisation and ideologies and to penetrate to the core of individual and social reality" (p. 173).

Fromm's expression, "...the core of individual and social reality", suggests that he believed in the existence of a "human nature". Fromm based this opinion on two grounds: one "morphological" and the other "existential". According to the morphological rationale, "The main argument in favour of the assumption of the existence of a human nature is that we can define the essence of Homo sapiens in morphological, anatomical, physiological and neurological terms. ... Surely we must assume, unless we regress to a view that sees body and mind as separate realms, that the species man must be definable mentally as well as physically."[45] Much as he refers to the morphological rationale as the "main evidence", the chief rationale behind Fromm's view that a human nature exists is *existential.* Fromm's *most original contribution* is here. According to Fromm, what defines humans as a species is that they form a fundamental bond with the world without any map or direction sketched by instinct. Fromm writes: "Man is the only animal who does not feel at home in nature, who can feel evicted from paradise, the only animal for whom his existence is a problem that he has to solve and from which he cannot escape."[46] It is this existential conflict that determines the nature of the species. This conflict "...produces certain psychic needs common to all men. He is forced to overcome the horror of separateness, of powerlessness, and of lostness, and find new forms of relating himself to the world to enable him to feel at home."[47] Although the need felt for these new forms is common to the human species, every culture meets the need in different ways.

Fromm's addition of the "existential dimension" to the definition of the human species is a valuable contribution that can fill a major gap in the Marxist approach to the human psyche. As we know, the materialist view of history explains all societal life; it therefore also explains the intellectual products of the society through the activity of the reproduction of the material base of societal life. However, there is a problem here. The content of knowledge and the existence of ignorance can be explained in this way, but it is not possible to explain the *contents of ignorance* in this way. Engels clearly states this: "Now as regards the more rarefied ideological fields, such as religion, philosophy, etc; these have a prehistorical fund of what today would be termed rubbish which was taken over lock stock and barrel by the historical period. In so far as these various false conceptions of nature, of the nature of man, of spirits, magic forces, etc, are economically based, it is only in a negative sense; false conceptions of nature are the corollary of the low level of economic development in the prehistorical period, but also on occasion its precondition if not its actual cause. And even if economic necessity may have provided the main incentive for progress in natural science and done so to an increasing extent, it would be pedantic to seek economic causes for all this primitive rubbish."[48] As seen, in his explanation of the primitive thought-belief world, Engels argues that economic elements can only have a "negative" effect, i.e. can explain the existence of ignorance; but in contrast, he clearly states, "...it would be pedantic to seek economic causes for all this primitive rubbish". In that case, can an explanation be given for this "primitive rubbish" and the more evolved forms it later took, without departing from the materialist conception of history? The answer to this very important question can be found in Fromm's "existential" dimension of the human species. In my view, Fromm correctly states that humans are the only living creatures on earth with the "ability to comprehend" and the only animal for whom "his existence is a problem that he has to solve and from which he cannot escape". Humans cannot feel secure unless they can find-invent this reassuring explanation. This explains the need underlying the intellectual conception of the myriad of creation myths, religions, and beliefs in supernatural powers that have existed since ancient times. It seems evident to me that, in prehistory and throughout history, belief-based ideologies have been continually rewritten in response to economic conditions. The materialist conception of history provides us with the intellectual tools to comprehend this mechanism. However, the best explanation of the "existential need" at the root of all these ideologies, in my view, belongs to Fromm.

According to Fromm, to understand the individual we must start with society. This belief was based on the fact that the individual character originates from inter-individual relations, as argued by Sullivan, and not from their biological, especially sexual, impulses as suggested by Freud.[49] Naturally, if the psyche originates from relations, then society itself must be examined. Which type of relations will be dominant in any society is determined by the society's mode of production, or put more generally, by its economic-social foundation and the accompanying mass of political and intellectual structures. Although all societies generally have numerous relationship forms, there is always a certain type that particularly fits with its own fundamental character. If we determine this type, we can also determine the "social character" specific to that society.[50]

Fromm focuses on this issue, particularly in *Man For Himself*, published in 1947, researching which types of society create which kinds of social characters. His studies in this respect are extremely impressive, particularly the social character created by the conditions of capitalist society after World War II.[51] In this society, the individual sees themselves as goods seeking a buyer in a market. Therefore, everyone's value in capitalist society is not their own use value (attributes), but their exchange value, just as with goods. The important thing is to have as high a value as possible in the personality market and to use it to find buyers. The notion of seeing ability to "sell" oneself as the most important characteristic is instilled in every individual by the family, education, etc., even when they are still growing up. The individual's self-view is shaped by this thought process. Whatever their individual talents, individuals only have self-value if they can find a buyer for themselves. As everyone cannot be "successful", the feeling of seeing oneself as worthless is alarmingly widespread in this kind of society, i.e.: "It makes no difference what my personal qualities are. If I have failed to find a buyer, it means that I am worthless." In this kind of society, equality means everyone in a certain field of activity can be replaced by someone else. Such an equality is the opposite of individuality. According to this, the condition for having value as a human, bearing in mind that value is the ability to meet demand, is not to have this or that quality; on the contrary, it is best to have no definitive characteristic at all in order to conform to a demand that could change any minute according to fashion or market needs. Thus, the empty space can be filled with whichever quality is in demand at the time. The attitude of the individual towards property, thought, and other people is always shaped by this concern, for example, "What should I eat and wear, what kind of house should I live in, what kind of car shall I drive, what type of thoughts and opinions should I identify with and who shall I get to know in order to have the highest value in the personalities market?"[52]

According to Fromm, for this social determination to yield results, one important condition is that individuals are not aware of the situation they are in. Such a system could not be sustained if everyone clearly saw that they had turned into goods produced for the market. Some mechanisms are therefore necessary to prevent the blossoming of this consciousness. Fromm calls this the "social subconscious".[53] This subconscious inhibits us from seeing that everything is done for money. A constant brainwashing convinces us that freedom, real equality of opportunity, and self-esteem can only be achieved in this kind of society. Thoughts, attitudes, and movements that question this idea are either ignored or prevented from entering social consciousness through a number of means. If we are to apply Fromm's observation to today's world, we can see some of the major problems that originate from the capitalist mode of production. For example, the fact that despite there being enough products to feed the entire world, over a billion people are continually undernourished, or the fact that global warming is an inevitable result of a consumer society. Yet somehow these realities cannot make their way into people's consciousness, because at the same time, politicians, media, universities, etc., do their utmost to prevent such realities travelling from the social subconscious into the conscious.

Provided that they stay within this common social personality framework, every individual acquires a distinctive, separate character, not shared by anyone else. This personality has two main origins. One origin consists of the traits acquired by everyone through heredity. Temperament has an important place among these. Different people may have a cheerful, sad, dynamic, or lethargic disposition. These traits do not determine what the person will do, but rather *how* they will do it, in other words, their mode of behaviour. The other origin consists of the personal relations that everyone establishes from the minute they are born. Part of this is unconscious, and it is where the facts of the social unconscious are kept. Without even noticing it, the individual suppresses their subconscious due to the fear of loneliness and isolation from society.[54] This is the basic driver of behaviour. As these social relations are determined by the society in which they occur, the personality acquired in this way basically shares the same qualities as the social personality. However, no two individuals, not even siblings, have exactly the same heredity and life experience. The individual tendencies resulting from these different datasets create a unique personality in every individual.

Freud states that by raising their consciousness, individuals can be saved from the captivity of their own subconscious. According to Fromm, an individual who achieves this, who is freed from being a toy of rationalisation, can also pull back the curtain of ideology concealing the social unconscious. This is enabled by the fact that individuals can never be entirely shaped by society. If people could be infinitely moulded, says Fromm, social institutions and rules that are harmful to individuals would last forever. When we look at history, we can see that people have never been entirely reduced to the status of puppets, and that when the time came, they were able to find the strength to change social patterns.[55] Fromm believed that the explanation for this should be sought in human nature. Although human nature is variable, its basic distinguishing characteristic is not. This characteristic is defined as follows: "In the art of living, man is both the artist and the object of his art; he is the sculptor and the marble; both the physician and the patient."[56] In short, whatever the characteristics of their society, humans are creatures that can ultimately pull the strings of fate with their own hands.

As can be easily seen from this short summary, Fromm's theory largely exposes the connections that historical materialism tries to establish between the society-individual poles, also revealing the mediations it seeks to illuminate. Humankind is its own architect. While the two approaches are not completely identical, there is nevertheless a great similarity between Fromm's understanding of human nature and Marx, who said that humans were both the playwright and the players in their own play.[57]

6. The Theory of the Individual According to Historical Materialism and Pierre Bourdieu

Bourdieu (1930–2002) was one of the most important sociologists of the twentieth century. Therefore, it is interesting to look at how his work might contribute to the historical materialist understanding of the individual. Bourdieu not only developed a theory of societal life but also applied it to numerous different areas of society. It is not possible here to summarise or give a general evaluation of the rich body of work he left behind. But it will be sufficient, for the purposes of our subject, to summarise the main threads of Bourdieu's general theory and evaluate it in the framework of historical materialism.

According to Bourdieu, as every society develops, numerous autonomous relationship networks form, increasing in number in parallel with the division of labour. Each has its own dominant values, legitimacies, norms, and behavioural rules. Some examples of these autonomous networks are education, sport, the media, academia, literature, and branches of the arts. Bourdieu refers to each one of these networks of relations, which each form a separate universe, as "social fields". The most distinctive feature of these separate spaces is that they all contain their own exclusive differentiations, i.e. they are based on a foundation of inequality.

Some of those who inhabit these spaces are closer to economic, cultural, or symbolic possibilities than the rest. In other words, they are in a superior position to others when it comes to taking advantage of the power centres that Bourdieu calls "capital". The basic differentiation dividing up each space originates from here. Additionally, there is no homogenous intra-class composition within the economic (based on financial power), cultural (based on endowed skills), or symbolic (based on respect for position) upper/lower class differentiations. In terms of their potential to access the power centres, the upper and lower class are split into many layers and, moreover, many individual positions. Ultimately, as a result of power differentiation, every social space becomes a constant battleground for maintaining or increasing power.

Bourdieu's originality lies in his observation that the individuals in these various social battlegrounds enter the relevant space with a pre-acquired configuration. Bourdieu calls this "habitus". The individuals of every field participate in society by unwittingly acquiring the world view, values, and rules of the behaviour that is specific to that field; this occurs through an internalisation that begins at birth. They see these particular world views, values, and rules of behaviour as universal truths. However, in reality, these truths relate to an order based on differentiated power and fundamental inequality. Thus, every order has the opportunity to portray the inequality among its own members as "natural" and "legitimate". Each person in each field "knows their place" and behaves accordingly.

Every social field has its own specific legitimacies. These legitimacies originate from the norms of the relevant space and the battles waged by various societal forces in their own interests. In these battles, words and symbols are a completely separate force. For example, it is derogatory to equate youth with "inexperience", but laudatory to talk of "boldness"; it is laudatory to equate old age with "wisdom", but derogatory to equate it with "decline". As can be seen, words or symbols are immensely powerful weapons in this societal battle. Bourdieu calls this effort to legitimise a discriminatory, hegemonic order as "symbolic violence", describing one of the primary tasks of sociology as unmasking these symbolic deceptions.

According to Bourdieu, in today's society, the main determinant in the establishment of the symbolic hierarchy is the state. The state imposes a symbolic order on the whole society. Thus, it shapes social perception patterns, common worldviews, common understanding, and social memory. Consequently, it is easy to understand why various power centres in society try to exert their influence onto the state. The reality of societal life is determined by words and symbols. The chief determinant of words and symbols is state power.

No social field or order is static. On the contrary, objective differentiation turns every order into a battlefield. Classes, strata, and individuals wage a constant struggle to maintain or improve their relative position to the centres of power. The struggle of every new generation benefits and builds on the possibilities created by the previous one. In other words, the people in a certain field have varying degrees of potential to transcend the borders erected by their "habitus". The gradual expansion of this potential may one day lead to a fundamental change in the balance of power and the formation of new habitus in the field concerned.

What contribution can Bourdieu's general theory, as roughly summarised above, make to a historical materialist theory of the individual? Firstly, it should be remembered that the phenomenon Bourdieu named "habitus" was also embraced by Marx. It is certainly now worth repeating some earlier observations I wrote in relation to this: "A social individual is *always an individual already determined.* As Marx and Engels state in *The German Ideology*, 'The fact is, therefore, that definite individuals who are productively active in a definite way enter into these definite social and political relations.'"[58] As seen, the distinctive feature of Marxism is that it deals with *all* variables as *pre-determined:* determined humans in a determined style with determined ways (style and ways create the form) enter determined relations. Bourdieu's additional contribution came from his valuable and meticulous examination of how habitus are determined. His most famous work on this subject, *La Distinction*,[59] published in France in 1979, showed how personal taste is formed by social position.

The research method developed by Bourdieu is very helpful for concretising the third dialectic identicality established by historical materialism between the individual and society. In the first chapter, I stated that to clarify this identicality it would be necessary to establish the *mediations* through which the individual-society poles transform into each other. Bourdieu's "social field" and "habitus" theory, together with the research he did to concretise this theory, are valuable guides that illuminate the determination of these mediations. With this type of research, it can be concretely determined how an individual's social position in a particular cultural framework determines their feeltypes and to what effect. As Bourdieu demonstrated, for an individual that is part of a certain culture, it is primarily their social position that determines how and what they will like, who they will trust and what they will fear most, what they will approve of and what will create shame, how they will experience their joy and sadness, what will give them pleasure or pain, and especially what they will be curious about and what they will struggle to understand.

As shown, Bourdieu's work makes a valuable contribution to historical materialism's theory of the individual, especially the understanding of its *intermediate levels*. However, while benefiting from this contribution, there are a few points where Marxists need to exercise caution. For one thing, Bourdieu's use of the word "class" is not the same as the concept of class adopted by Marx. Therefore, in the framework of historical materialism, the social reality that Bourdieu referred to as "class" should be renamed as something like "strata" or "groups". The concept of "capital" used by Bourdieu is also not the same as the sense in which it was used by Marx. It should be called something different, such as "power centre". Thirdly and most importantly, unlike Marx, Bourdieu failed to acknowledge that the social influence originating from production relations was the chief determinant of power, seeing cultural and symbolic possibilities as equal or even sometimes of more importance. Much of the time, Bourdieu saw symbolic reality as the fundamental social reality. Whereas for Marx, the main determinant of symbolic reality was the concrete mode of production of the relevant society.

It should also be noted that Bourdieu's theory says nothing about the "individual" reality of the individual-society dialectical identity that historical materialism aims to understand. Therefore, even if the historical materialist view includes Bourdieu's work with the above provisos, it will still not succeed in the task of constructing its own specific theory of the individual in its own theoretical structure. For this, there must also be a theoretical explanation for the reality of the psyche. In my opinion, the feeltype theory elaborated in this work is a vehicle that will help to fill the gap.

7. Cultural Psychology and Feeltypes

Of the many different scientific approaches to culture and psyche that have arisen from the USA in recent years, *cultural psychology*[60] comes closest to the ideas of psyche-culture interaction presented in this book. The definition of cultural psychology is as follows: "Alternatively stated, cultural psychology is the study of all the things members of different communities mentally experience (know, think, want, feel, value), and hence do, by virtue of being the kinds of beings who are the beneficiaries, guardians, and active perpetuators of a particular cultural tradition."[61] Within this approach, the attributes of the individual psyche (knowledge, thoughts, desires, feelings, values), along with the behaviours that result from their influence, are explained by the fact that they are part of a certain cultural tradition, which the individual benefits from, guards, and actively perpetuates. This approach argues that culture and psyche are always *intertwined* when studying the psychical foundations of culture and the cultural foundations of the psyche. Wherever there is a psyche, there will undoubtedly be a culture that shapes and develops it, enabling it to exist. Likewise, wherever there is a culture, there will undoubtedly be psyches that make it concretely exist, and that develop it together with the behaviours that are their outer manifestation. Evidently, this approach is very close to the viewpoint presented in this work. (One of the reasons I do not say it is identical is that within the same cultural psychology framework there are diverse approaches, touching on different aspects, therefore throwing up some discrepancies. Consequently, it is not possible to talk of a completely homogenous cultural psychology that could be part of this identicality.)

One of the basic problematics of cultural psychology is the question of whether humans share some universally valid characteristics beyond all their specificities. According to one viewpoint, human communities, and hence their members, are all specific to themselves. There is no essence or "nature" belonging to the human species.[62] At first glance, this seems to conform with the portrayal of human nature provided by Marx in his *Thesis on Feuerbach VI*, where he writes that human nature "is the ensemble of human relations".[63] However, the interpretation of endless variability is not actually compatible with Marx's view. Furthermore, as it is one-sided, it is contrary to a dialectical outlook. The Marxist expression of "the ensemble of human relations" cannot be separated from the view of human nature based on the human act of "constant reproduction of existence, relations and tools". Therefore, universality exists, but it always materialises in the specificities of a concrete society (lived or living in a certain space and time). These specificities are not entirely random or coincidental. If it is a "human" community, i.e. a "society", that we are looking at, the activities connected to the reproduction of the society's material existence and basic relations will undoubtedly be the most significant of all the relations established between individuals.

A similar universality-specificity approach is argued within the framework of psyche-culture relations in the article used as the main source here. This is clearly reflected by the wording of the article heading, "One

Mind, Many Mentalities". In other words, "one attitude, many habits" or "one intellect, many perceptions". This approach can be of great help in understanding the psyche–culture relationship. Universality should be sought in the common existence of *possibilities*. However, these possibilities emerge in the form of different symbolic structures and behaviour patterns in every culture.

In cultural psychology, the main subject of investigation is particular traditional behaviour patterns. (In the article, this is referred to as a "custom complex".) Numerous examples of this custom complex can be cited: whether young children are put to bed with both parents, with just their mother, or alone; whether children are allowed to suckle whenever they like or if this is prohibited after a particular age; whether couples eat together or males eat separately; whether children's education encourages independence or social solidarity; etc. The first step towards investigating a custom complex is to observe whether such a traditional behaviour pattern exists in a certain space. The second step is to identify the associated information, thoughts, feelings, wishes, and values of those who are involved in it. Afterwards, the mentality (understanding, ethos) upholding this psychical structuring, which leads to a certain pattern of practice, can be identified.[64]

There is no consistency bias in the cultural psychology approach. In other words, the same mentality might not necessarily be dominant in all the traditional custom complexes of a certain culture. On the other hand, there is a possibility that it could be the same and this subject also needs to be studied. For example, as repeatedly stressed in the article I use here as a main source, according to recent research, a fundamental difference in ethos exists between North American and Western European societies on one side and Eastern societies on the other, particularly those like China, Japan, and Korea. This difference is manifested in many different fields. While mutual dependency, solidarity, and collectivism are dominant in the custom complexes of the East, independence and personal autonomy, in other words individualism, prevail in the West. In the custom complexes of the East and West, behaviours considered to be right, good, and normal are complete opposites in the two cultural worlds. Naturally, it is unrealistic to seek absoluteness in this determination, as neither East nor West consist of totally homogenous cultures. It is nevertheless interesting that so many different studies have revealed such a difference in emphasis.

The home environment in which children traditionally grow up has a determinative role in instilling cultural customs. There are many aspects that could be examined when studying these environments. For example, are the home and work environment intertwined or completely separate? Is the home environment made up of a large or small number of individuals? Does it contain just mother and father and similar close relations or, as in the case of tribes, are more distant relations also considered part of the home? Are the relations in the home environment regular or sporadic? Is the home environment open or closed to other social environments? Is childcare predominantly down to the mother or is it a shared responsibility? Is the parent–child relationship child- or adult-centric? Are children treated differently according to their age or treated the same regardless? Etc. After identifying these and other similar traditional custom complexes, the second step of analysis is to uncover the perspective (world view) behind each one.

Another important area of cultural psychology is the role of language in a child's adoption of traditional symbols and behaviours. Every child learns a language, but languages differ. Every language expresses the symbolic world and behaviour patterns of a certain culture, and it is also the most effective tool for transferring them. Thus, as a child learns a language, they simultaneously internalise the perspectives and values that govern the dominant symbolisations and custom complexes of their own culture. In this way, language is more than just a framer of meaning, it is also a *framer of identity*.

The impact of culture on an individual's self-perception and the formation of their personality, or ego, is one of the most interesting subjects in cultural psychology discourse. To identify this, the perception of "self" specific to each culture must be researched. In a particular culture, how is the question of "what am I" or "who am I" posed? Is it an important question or relatively insignificant? When answering the question, to what degree are the dominant values of the culture used as criteria in the definition of self? What perception stands behind such an attitude of "being me"? Hence, with this last research subject, cultural psychology fulfils the most fundamental demand of its approach, which argues that psyche and culture create each other and cannot exist apart.

To what extent does this correlate with the approach put forward in this book? The main issue is the subject of universality-specificity. Both approaches believe that universality can only come about in the form of specificities. However, when it comes to the principal elements of universality, there is a divergence. According to cultural psychologists, the common elements consist of the following: whichever culture humans are part of, they have the abilities to acquire knowledge, think, feel, want, and value, which allow them to become familiar with their environment and predetermine their own behaviour. The common aspect of all cultures is that they are based on individuals endowed with these abilities. When we go beyond that to the subject of what will in concrete be known, thought, felt, wanted, or valued, there is no longer any universal aspect among people. The concreteness is always established by cultures in their own specific way.

In my view, there is no real possibility of finding any common aspect among cultures when universality is defined so broadly, because the psychical abilities presented as a commonality are actually empty shells. This means nothing more than saying, "The common aspect of all cultures is that they are limited by being human cultures". The feeltype theory argued here allows this empty shell to be filled. According to the feeltype theory, just like human bodies, human psyches are equipped with some universal sensibility axes. Just as the human body knows the outside world through a limited number of sensibility axes, so does the psyche know the outside world through a limited number of sensibility axes. The body has five of these, which we call senses: sight, hearing, smell, taste, touch. The psyche has six, and we call these "feeltypes": Like/Dislike, Trust/Distrust, Approval/Disapproval, Happiness/Sadness, Pleasure/Pain, Comprehension/Confusion. The objects that the senses concretely sense are dictated by the environment, as are the objects felt by the feeltypes. In both cases, this "environment" is culture.

Clearly, the feeltype theory is ideal for examining traditional custom complexes like those mentioned above, and these are the questions that must be asked for each behaviour pattern: for the individuals involved in these complex patterns of relations, what are the objects of like and disgust? Who and what are trusted, who and what are feared? Which behaviour is approved, and which brings shame? What brings happiness and what brings sadness? Which people and behaviours give pleasure? Which cause pain? Is the bundled mass of relations that make up the complex pattern understood by everyone who plays a role in it? Otherwise, is it difficult to comprehend? And lastly, in which ways are all these feelings expressed?

The feeltype theory can also facilitate understanding of the socialisation-individuation that occurs during the growing process of young individuals. The immediate environment of the child, the "home environment" as mentioned above, can be best analysed by examining the affection, trust, approval, joy, pleasure given to the child, along with the possibilities for them to understand their environment and themselves.

The feeltype theory is also useful for analysing what effect learning to speak has on the establishment of the psyche. Structures, fictions, and symbols, or in other words, forms of concrete manifestation, exist in the concreteness of every culture and every language. They show children how and what will be liked, how and what will be trusted, what will be approved, the reasons for happiness, the sources of pleasure, and how they will understand a subject. These facilitate the establishment of a child's psyche in a form that is compatible with the culture in which they live.

In short, the feeltype theory is functional for all the major study areas of cultural psychology. It is therefore also of particular use for *comparative* studies.

Notes

1. Connerton (1989, p. 74).
2. Kâğıtçıbaşı (2010, pp. 379–380).
3. Özbek (2010, p. 153).
4. And they repaid us. To understand that, you only have to look at the various types of zoonotic flu (bird flu, swine flu, etc.) or AIDS (originated from monkeys) that we have not yet been able to overcome.
5. For example, as seen earlier (Chapter II, section "Lev Vygotsky's Cultural-historical Psychology Approach"), for Vygotsky, there is no such problem as socialisation. The problem that needs to be analysed is the *individuation* process of the human juvenile coming into the world as the member of a species.
6. Rose (1996).
7. Kardiner (1945, p. 112).
8. "The meanings that symbols, the material vehicles of thought, embody are often elusive, vague, fluctuating, and convoluted, but, they are, in principle, as capable of being discovered through systematic empirical investigation – especially if the people who perceive them will cooperate a little – as the atomic weight of hydrogen or the function of the adrenal glands." Geertz (1973, pp. 362–363).
9. This mistake is made by Lucien Sève. For more on this subject, see Eroğul (2007, pp. 266–267).
10. In my view, the two most valuable studies of the class structure in Turkey are those of Korkut Boratav listed in the References section.
11. The basic information about human evolution is from Özbek (2010). The citations are also from there. Any additional comments are mine.

12. Marx (1976, p. 284).
13. http://fr.sott.net/articles/show/4608-Neandertal-et-Cro.Magnon-auraientmelange-leurs-genes. Accessed 20 August 2011. And, http://www.guardian.co.uk/science/blog/2011/aug/25/neanderthal-denisovan-genes-human-immunity. Accessed 26 August 2011.
14. Engels (1876).
15. Marx and Engels (1998, p. 37).
16. "As culture shaped us as a single species – and is no doubt still shaping us – so too it shapes us as separate individuals." Geertz (1973, p. 52).
17. Eroğul (1981, p. 56).
18. "Human thought is consummately social: social in its origins, social in its functions, social in its forms, social in its applications." Geertz (1973, p. 360).
19. Mead (1972).
20. "It is absurd to look at the mind simply from the standpoint of the individual human organism; for, although it has its focus there, it is essentially a social phenomenon; even its biological functions are primarily social." Mead (1972, p. 133).
21. Mead (1972, p. 112).
22. Ibid. (p. 154).
23. See Chapter II, section "Lev Vygotsky's Cultural-historical Psychology Approach").
24. Holland and Lachicotte, Jr. (2007, p. 105).
25. Or in his own words: "what distinguishes the worst architect from the best of bees is that the architect builds the cell in his mind before he constructs it in wax." Marx (1976, p. 284).
26. Mead (1972): 173.
27. *The Neurotic Personality of our Time.* In this work, Horney mentions those who emphasised the importance of culture in the formation of the psyche, in contrast to Freud's biological approach. She sincerely credits the pioneers as Erich Fromm, in the field of German psychoanalysis literature, and Harry Stack Sullivan in the USA. Horney (1937, p. 20, n. 10).
28. Bowlby refers to this connection that forms the main bridge between our lives as "attachment". Similarly, according to Erikson, a "basic trust" develops in everyone, also due to this connection.
29. *Neurosis and Human Growth.* The publisher acquired the copyright of this work in 1950, but it was not published until 1970, long after Horney's death. The references here relate to its latest edition, published in 1991 with a new preface.
30. The psychoanalyst R. D. Laing (1927–1989) provides one of the best analyses of the workings of the second "fictional" personality in schizophrenia-type neuroses. *See* (1990): *passim*.

31. Evans III (1996, p. 2).
32. Sullivan (1964, pp. 210, 220–221, 224).
33. Ibid. (p. 225).
34. Evans III (1996, p. 113).
35. Chapter II, section "Feeltypes and Piaget's Theory of Soul Formation"), with reference to Piaget.
36. The importance and quality of the emotional bond established with the mother, or another person in the mother's role, for the emergence and establishment of feelings of trust in the human juvenile has been examined in most detail by John Bowlby. Therefore, Bowlby's work will be dealt with in a separate subsection.
37. Summarised and cited by Evans III (1996, pp. 80ff.).
38. Holmes (2011, p. 2).
39. While behaviourism focuses on behavioural reactions in a laboratory environment, ethology studies the behaviour developed by animals in a natural environment.
40. Bowlby (1988). All the page numbers given in parentheses in this subsection refer to this work.
41. As can be seen, Bowlby provides a powerful theoretical explanation that helps us to understand assertions like "I killed them because I loved them."
42. His main work *Attachment and Loss*, published in three volumes between 1969 and 1980 runs to around 1500 pages.
43. Holmes (2011, p. 51).
44. Ibid. (p. 56).
45. Fromm (1973, p. 220). As shown, the rationale Fromm used to defend the morphological identicality of the intellect in the human species is that the body and intellect are inseparable. In other words, he can only defend morphological identicality of the intellect by alluding to somatic identicality. Whereas the feeltype theory I present in this book provides a direct basis for defending the morphological identicality of the intellect of the human species.
46. Fromm (1973, p. 225).
47. Ibid. (p. 226).
48. Engels, F. Letter to C. Schmidt, 27 October 1890 (pp. 687–688).
49. Fromm writes, "This concept was very adequately formulated by H. S. Sullivan in defining psychoanalysis as a 'study in interpersonal relations.' Interpersonal relationships, assumed by Freud to be the result of varying forms of sexual desire, are considered to be the factor determining sexual strivings. In this view it is not sexual behaviour that determines character, but character that determines sexual behaviour." Fromm (1948, p. 2).

50. Abram Kardiner later used ethnographic data to develop Fromm's notion of "social character" in a conceptual framework called "basic personality". Eroğul (2007, pp. 235–236).
51. Fromm (2003, pp. 50–57).
52. The worry of adjusting your whole life to the market has today even reached the stage of the "commodification of feelings". Hochschild (2003), *passim*.
53. Fromm pays particular attention to the social subconscious in his book *Beyond the Chains of Illusion* (1962).
54. "But there is still another and, as I believe, the most powerful motive for repression: the *fear of isolation and ostracism*." Fromm (2009, p. 94).
55. Fromm (2003, p. 15).
56. Ibid. (p. 12).
57. Marx (1963, p. 115).
58. Marx and Engels (1998, p. 41).
59. *Distinction: A Social Critique of the Judgement of Taste* (English Trans. Routledge. 2010).
60. Çiğdem Kâğıtçıbaşı, who is sadly no longer with us, was one of the greatest advocates of this approach in the Turkish academic world. *See* Kâğıtçıbaşı (2000).
61. Shweder et al. (2006, p. 719).
62. Ibid. (p. 723).
63. Marx (1845, p. 29).
64. *Handbook of Child Psychology, Vol. I.* (2006, p. 727).

References

Bowlby, J. (1988). *A Secure Base Parent-Child Attachment and Healthy Development*. Basic Books.

Connerton, P. (1989). *How Societies Remember*. Cambridge University Press.

Engels, F. (1876). The Part Played by Labour in the Transition From Ape to Man. In *Karl Marx and Frederick Engels Selected Works in One Volume* (pp. 354–364). Lawrence and Wishart, 1970.

Eroğul, C. (1981). *An Essay on the Nature of the State*. University of Ankara, Faculty of Political Sciences Publications.

Eroğul, C. (2007). İnsanın Var Olma Biçimi Olarak Birey [The Individual as Form of Existence of Humankind]. 10. Ulusal Sosyal Bilimler Kongresi Metinleri [Papers submitted to the 10th Social Sciences Conference]. In *Türkiye'de Neoliberalizm* (pp. 235–272). Yordam Kitap, 2009.

Evans, F. B., III. (1996). *Harry Stack Sullivan, Interpersonal Theory and Psychotherapy*. Routledge.

Fromm, E. (1948). Sex and Character: The Kinsey Report Viewed From The Standpoint of Psychoanalysis. First published in D. P. Geddes and E. Curie (Eds.): *About the Kinsey Report* (pp. 301–311). The New American Library. [Provided by the Fromm estate.]
Fromm, E. (1973). *The Anatomy of Human Destructiveness*. Holt, Rinehart and Winston.
Fromm, E. (2003). *Man For Himself*. Routledge.
Fromm, E. (2009). *Beyond the Chains of Illusion, My Encounter with Marx and Freud*. Continuum.
Geertz, C. (1973). *The Interpretation of Cultures*. Basic Books.
Hochschild, A. R. (2003). *The Managed Heart, Commercialization of Human Feeling*. University of California Press.
Holland, D., & Lachicotte, W., Jr. (2007). Vygotsky, Mead, and the New Sociocultural Studies of Identity. In H. Daniels, M. Cole, & J. V. Wertsch (Eds.), *The Cambridge Companion to Vygotsky* (pp. 101–135). Cambridge University Press.
Holmes, J. (2011). *John Bowlby and Attachment Theory*. Routledge.
Horney, K. (1937). *The Neurotic Personality of Our Time*. W. W. Norton & Company, Inc.
Horney, K. (1991). *Neurosis and Human Growth*. W. W. Norton & Company, Inc.
http://fr.sott.net/articles/show/4608-Neandertal-et-Cro.Magnonauraient melange-leurs-genes. Accessed 20 Aug 2011.
http://www.guardian.co.uk/science/blog/2011/aug/25/neanderthaldenisovan-genes-human-immunity. Accessed 26 Aug 2011.
Kâğıtçıbaşı, Ç. (2000). *Kültürel Psikoloji. [Cultural Psychology]*. Evrim.
Kâğıtçıbaşı, Ç. (2010). *Günümüzde İnsan ve İnsanlar [Man and People Today]*. Evrim.
Kardiner, A. (1945). The Concept of Basic Personality Structure as an Operational Tool in the Social Sciences. In R. Linton (Ed.), *The Science of Man in the World Crisis* (pp. 107–122). Columbia University Press.
Laing, R. D. (1990). *The Divided Self*. Penguin Books.
Marx, K. (1845). Theses on Feuerbach. In *Karl Marx and Frederick Engels Selected Works in One Volume* (pp. 28–30). Lawrence and Wishart, 1970.
Marx, K. (1963). *The Poverty of Philosophy*. International Publishers.
Marx, K. (1976). *Capital Volume One*. Penguin Books in association with New Left Review.
Marx, K., & Engels, F. (1998). *The German Ideology*. Prometheus Books.
Mead, G. H. (1972). *Mind, Self, and Society*. (C. W. Morris, Ed. and Introduction.). The University of Chicago Press.
Özbek, M. (2010). *İnsanın Tarihöncesi Evrimi [The Prehistoric Evolution of Man]*. Bilim ve Gelecek.

Rose, N. (1996). *Inventing Our Selves: Psychology, Power and Personhood.* Cambridge University Press.
Shweder, R. A., et al. (2006). The Cultural Psychology of Development: one Mind, Many Mentalities. In R. M. Lerner (Ed.), *Handbook of Child Psychology, Vol. I* (6th ed., pp. 716–792). John Wiley & Sons, Inc.
Sullivan, H. S. (1964). *The Fusion of Psychiatry and Social Science.* W. W. Norton & Company, Inc.

Conclusion

The main conclusions reached by this research, which aims to advance a Marxist understanding of the individual, are as follows.

Humans are not a type of animal. They are a unique living species. Humankind is the only social species. While there are many creatures that live in communities, living in a society is exclusive to humans. Society is the world of symbols, objects, and traditions in which the individual exists. The individuals who are part of this cultural world have no equivalent anywhere in the world. They are unique creatures, who imagine their environment in their brains, behave knowingly of their own volition, are able to carry out an aim visualised in their mind by first planning it in their head, and can constantly reproduce and change their material and spiritual worlds through the social relations in which they are involved.

The sociality that characterises humankind is the major variable explaining both the body and psyche. A body that is primed for the imprinting of a human psyche, in other words the individual form of society, is fundamentally different to any animal body. In terms of defining humankind, the biological commonality with animals is secondary. What makes humans human is that individual organisms have acquired, through evolution, the qualities enabling them to be part of a network of social relations. For humans to be capable of reproducing and changing their own material and spiritual life inside a network of relations with others, it was necessary for the front feet to become hands, for the larynx to evolve in such a way as to enable speech, and for the brain to reach the volume and shape able to

C. Eroğul, *Marxism and the Individual*,
https://doi.org/10.1007/978-3-030-83662-7_5

manage these relations. In other words, a prerequisite was the emergence of a body that could concretise the conventions constituting the world of culture in the form of individual habits.

What makes the human body the human body is not a set of biological characteristics shared with other near or distant creatures; it is that it is an organism that has acquired the qualities to convey sociality. Humans can no more be defined by the atoms, molecules, cells, and tissues they are made of than they can be by their genetic makeup, over 95% of which is shared by chimpanzees. Human biology, with its uniquely human characteristics, can only be explained as a product of sociality.

Not just the human body, but also the psyche, i.e. the individual feeling-thought world, originates from society. The materials of the psyche are the feeling-thoughts originating from inter-individual relations and the habits that concretise the individual's share of cultural conventions. Every psyche is a system. The feeling-thought materials that make up this system can only be carried by the biology specific to humans. Our habits are imprinted onto our body. Therefore, society is really in our body.

The feeling-thoughts that make up the human psyche are the product of a set of sensibilities that have been moulded by a certain culture. These sensibilities were created by an evolutionary process in which sociality was the determinant. Just as the human body perceives its environment through a set of sensibilities with five axes – the five senses – so does the human psyche sense its environment through a set of sensibilities with six axes – the six feeltypes. These are the like, trust, approval, happiness, pleasure, and comprehension feeltypes. Any feeling-thoughts that emerge from inter-individual relations in a particular culture are able to exist due to the sensitivities on these six feeltype axes: liking, being liked, disliking, being disliked, trusting, being trusted, fearing, being feared, approving, being approved, disapproving, being disapproved, feeling happy, making happy, feeling sad, making sad, having pleasure, giving pleasure, feeling pain, giving pain, understanding, being understood, not understanding, not being understood. Although their objects and the way they are experienced may differ between cultures, all these feeling-thought types are made possible by human sensibilities common to all cultures.

Cultural specificities can only be understood and explained if this commonality is seen. Every culture concretises these feeling-thought types through a specific world of symbols, a specific set of conventions. Conventions take the form of a series of roles, structures, and institutions, which differ in function and in their power to determine the individual.

These are brought to life by individuals-subjects using countless material and mental vehicles presented by every culture. The reproduction and transformation of material and mental life is executed by mechanisms specific to each culture. These mechanisms need to be researched and examined in order to understand and explain a particular culture. The toolset required for analysing social data has been provided, and continues to be provided, firstly by the materialist conception of history – especially Marx – and also by specialist social fields like sociology, anthropology, archaeology, cultural psychology, and evolutionary science.

The specificities of the individual psyche, and therefore of the body that carries it, can only be understood and explained if this common ground is also understood and explained. As part of a certain culture's common societal life, every individual shares some things with all the other individuals who are part of the same culture. However, as every individual life within this general shared framework is unique, every psyche is also unique. Moreover, as pointed out above, society itself could not exist without this relative difference. The reproduction and transformation of material and psychical life is only possible through the network of relations established between different individuals. Each individual has a unique experience of the culture-specific feeling-thoughts that emerge as a result of the six feeltypes being moulded by a certain culture. Every individual shapes these feeling-thoughts in their own unique way, experiencing them after they have passed through their own unique synthesis. This set of unique experiences builds a unique psyche for every individual. Through a unique process of ontogenesis, the individual acquires the habits that make them who they are. The building blocks of each individual psyche's system are the feeling-thoughts that are acquired as part of this process and are exclusive to the six feeltypes. Hence, understanding and explaining an individual means analysing their own unique system, a system which is largely unconscious.

The toolkit needed for this analysis has been provided, and continues to be provided, by psychoanalysis – especially Freud – and by psychiatry, psychology, biology, genetics, and the expertise of various other branches of medicine. In order to understand the individual, it is also necessary to understand and explain individual intelligence. The necessary toolkit for understanding and explaining the individual intellect has been provided, and continues to be provided, firstly by the founder of genetic epistemology – Piaget – and also by those working in the fields of neural research, informatics, and artificial intelligence.

The Marxist approach to the individual must absorb and rework all these contributions. The main point of consideration in this reworking is the dialectical identicality of the individual-society relation. There is no question that the specificities of sociality and individuality must be researched separately. However, the dialectical identicality of these two elements must not be overlooked at any stage of the research. The ultimate aim is to express all the specificities of sociality with individual-specific mediations and to express all the specificities of individuality with society-specific mediations. This huge task is not one that can be tackled quickly by taking shortcuts with the methodology. Nevertheless, these efforts are ongoing and there have been numerous contributions so far. In my opinion, the Marxist approach is the most promising way to join all these contributions together.

According to historical materialism, the most effective way to unravel the individual/society problematic is to tackle the problem from a historical point of view, or to use a more comprehensive term, a "genetic" point of view. As already stated, the origin of the individual/society phenomenon is evolution, the basis of Darwinism. We know that about 30,000–40,000 years ago, the evolutionary process ended with the dominance of the only human species left on earth today, *Homo sapiens*, represented by the Cro-Magnons. About two and a half million years ago, the genus *Homo*, the origin of this species, started to emerge when a group of hominids began to cooperate in reproducing the material basis of life. The material condition for this cooperation was using stone tools made with the aim of capitalising on nature. It is this phenomenon that explains the continuous development, during a time period of about two and a half million years, of the two most important organs that make humans human – the hands and brain. This view is probably the most fruitful scientific contribution to understanding human reality.

However, the collective production that necessitates inter-individual cooperation also depends on humans being able to understand each other. Cooperation necessitates communication. Communication necessitates a common communication medium; this is language. Spoken language depends on humans being able to attach common meanings to various sounds and perceive them as "symbols made up of sounds". This development was supported in the biological evolution process by the development of a larynx able to produce enough different sounds and of a brain able to grasp enough different meanings. This is the phenomenon that explains the formation of society-individual.

The greatest explanation of individual affectivity and ability to think, and therefore how consciousness stems from social communication, was put forward by George H. Mead. It is thanks to Mead that we understand how the individual is formed by societal roles determined by culture. The stages this formation passes through during a child's growing up process were best researched by Jean Piaget. However, both scholars overlooked the variability of these societal roles. Lev Vygotsky compensated for this oversight, stating that the societal relationship patterns that constitute the individual are connected to particular historical developments.

This is how the emergence and formation of the individual/society duality is explained. Yet, prior to Freud, the psychical reality of the individual aspect of this duality was not dealt with as an object of original scientific research. It was Sigmund Freud who made this gigantic leap. Freud's greatness stems from his discovery that the psyche is basically unconscious and made up of affectivity more than intellect. However, Freud emphasised the precedence of biological drives in the formation of the individual, consequently overlooking the sociality-historicity that is the real source of humankind. It was Harry Stack Sullivan who argued that the unconscious affective aspect emphasised by Freud was predominantly based on inter-individual affective relations. First Sullivan, and then John Bowlby, asserted that the more essential cooperation is for society to continue its existence, the more crucial it is to have a protective trust relation based on face-to-face relations, from birth onwards, in order to ensure that the individual has a more or less healthy psyche. Yet, both scholars overlooked the fact that these societal relationship patterns connected to fundamental security were first and foremost dependant on the culture of the relevant society.

The two most important scholars to argue that the development of the individual psyche is based on the cultural patterns of society are Erich Fromm and Karen Horney. Even today, no other person has been more successful than Fromm in usefully synthesising the ideas of Marx and Freud. Fromm stated that a dominant "individual character" model existed in every society, based on the dominant mode of production or the social order, and that the widespread adoption of this model is dependent on the "societal subconscious" concealing some social realities, such as exploitation. Fromm also argued, just like Marx, that the nature of the human psyche compels humans to "dominate their own destiny". Consequently, individuals can unveil the "societal subconscious" through common consciousness. Horney argued that the adaptation and defence mechanisms developed by individuals were basically determined by their culture and that the explanations Freud sought in biology should instead be sought in societal relations. Erik Erikson, one of the leading names in psychoanalysis, also developed a theory examining how personal development, which continues from birth to death, is linked to the dominant culture of the society.

Any psychological or sociological contribution that can be aligned with the fundamental fact that human individuals are social creatures should undoubtedly be used in the development of a Marxist theory of the individual/society. Bearing this in mind, the value of the behavioural/social learning theory based on the work of Ivan Pavlov, and later developed by John Watson, Albert Banderas, and others, should not be underestimated. This theory, with its reinforcement/punishment mechanisms, helps us to understand how social roles are willingly adopted by individuals. Even the biological approach persistently argued by Hans Eysenck contains a contribution to Marxism. It allows us to understand how the characteristics that define "temperament", such as extroversion-introversion or placidity-irritability, can be largely hereditary without denying the basic fact of individual sociality. The Big Five distinctive traits approach, originally developed by Gordon Allport and widely accepted in American psychology today, is a diagnostic and classification tool that is very useful for Marxists in the description of individual psyches. Additionally, the cultural psychology approach, which acknowledges the interconnectedness of culture and individual psychology, is also broadly compatible with a Marxist theory of individual/society. Finally, the "social fields" theory, which made Pierre Bourdieu one of the leading sociologists of our time, can be easily reconciled with a Marxist approach. The crucial point of this theory is that in every social field determined by the social order (and therefore basically by the mode of production and the culture), the individuals competing with each other for a share in the social power centres are, through their "habitus", already in fact determined by society. Like Marx, Bourdieu saw that the individual/society reality was in essence *synchronous* and *identical*. As such, there are vast possibilities for using the research tools developed by such a leading sociologist as Pierre Bourdieu to examine different segments of society from a Marxist standpoint.

The development of a Marxist theory of the individual has come to be of paramount importance. Communism is a major requirement for humankind to protect and develop its existence as a species. But that does not mean the type of repressive, statist socialism witnessed up until now.

In this communism, individual development and social development will be both the condition and the engine of each other. There is a need for an individual/society theory that can help to establish, explain, and develop such an order. In a letter he wrote in 1894, a year before his death, Engels stated that the dialectical understanding of history was "*bound* to be discovered", and that if Marx had not done so, someone else would have done it, because "the time was ripe" for it.[1] I think that now "the time has come" for a communist individual/society theory. I am glad to have been able to present but a small contribution to this gigantic task, which necessitates immense collaboration.

Note

1. Engels, F. Letter to W. Borgius, 25 January 1894 (p. 695).

Interview with Gökhan Atilgan

Interview start date: 15 July 2013

You began your work on the individual in 2002. That year marked a turning point in the history of Turkey; the Justice and Development Party (AKP) came to power, and subsequently most social scientists started to focus on issues like Islamism, neoliberalism, conservatism, and hegemony. You went against the grain and chose to focus on "the individual". If you do not object, I would like that to be the starting point of our conversation. Why did you start working on the issue of the individual?

Actually, there is no particular reason why I started in that year. I have been interested in the subject of the individual for a long time. In fact, my interest began at the age of fifteen, when I started taking a good look at myself and found that I did not like what I saw, in body or in mind. So, I started working out for half an hour every morning and carried that on without pause for thirty years. I also tried to educate myself and have not stopped since. After discovering Marxism at the age of twenty and opting for the field of "constitutional law and political science" in my academic career, I spent years researching the subject of politics and the state. After finishing my book, *An Essay on the Nature of the State*, I intended to look at "the capitalist state" and then "the state in Turkey". This process came to an abrupt end when I was kicked out of the University in 1983. When I returned seven years later, my first challenge was to produce material that could be used for lecture notes on "constitutional law" and "comparative government". The last of my two books on these subjects was published

C. Eroğul, *Marxism and the Individual*,
https://doi.org/10.1007/978-3-030-83662-7

in 1996. I was working on other unrelated subjects at the same time and that continued afterwards. As I got older, I felt the need to focus on a main subject again. I had already lost the thread on the subject of the state. One day it occurred to me that, as Marxism relates to every aspect of society, how can it have nothing to say about the individual? Somehow, the question I first asked myself at the age of fifteen, "Why am I like this?", turned into the search for a "Marxist theory of the individual". In short, the political climate of the time had no impact on my choice of subject. What drew me towards the subject more than anything was the need of a Marxist scholar, approaching sixty, to tackle his youthful quandary on a scientific level.

So, you began to research what Marxism had already said about the individual, and what else it might say. You realised that when it came to the individual, Marxism did not have a systematic, independent theory like the one developed for capitalism. Why is it that Marxism did not develop a systematic theory of the individual? Is that not odd? Is it because Marxists continued to focus on the areas most emphasised by Marx, even after he was gone?

Marxism is an extremely robust theoretical approach that can help with the study of every aspect of society. Societal life is such a rich subject that it can be the focus of pretty much countless studies. Since the main aim of Marxists is to transform society, it is not surprising that they have prioritised issues like the crisis of capitalism, imperialism, globalisation, monopoly, and the class struggle. Nevertheless, it should not be forgotten that, outside of these fields, Marxism contributed invaluable research on a wide variety of subjects, from art to ideology, and from literature to politics. There are also some valuable studies directly concerned with the individual. The first that spring to mind are: Lev Vygotsky's book *Thought and Language*, which was published in Russian in 1934; and there is Erich Fromm's book dating back to 1962, *Beyond the Chains of Illusion – My Encounter with Marx and Freud*, which tries to create a synthesis between Marx and Freud. Afterwards, Fromm continued to make important contributions to human theory, benefiting considerably from the Marxist perspective. In 1969, Lucien Sève, one of the leading thinkers of the French Communist Party, wrote a book called *Marxism and the Theory of Personality*. István Mészáros in 1970 and Bertell Ollman in 1971 carried out some extremely valuable research on alienation, clearly demonstrating the Marxist view of the human species. More recently, in his book *Bodies of Meaning*, published in 2001, David McNally strongly defended the Marxist understanding of the individual against postmodernism. Therefore, it would be wrong to say that Marxism has not produced any worthwhile thoughts about the individual until now. At the same time, it is true that Marxism has much more work to do on the subject. But this observation is equally valid for all areas of society. Considering that the third volume of *Capital* was never completed, and that capitalism has undergone radical changes since then, it cannot be argued, even concerning the theory of capitalism, that Marxism has nothing important left to say.

One work that is often mentioned on the subject is Adam Schaff's book, Marxism and the Human Individual, *first published in 1970. Yet, you do not mention it at all in your book. Can you explain why?*

I read Schaff's book as soon as it came out. It was a long time ago, so I am not in a position to give a detailed critique now. But I do remember that I found the book extremely poor on an academic level. It was written in opposition to the view that "Marxism places no value on humans and is averse to humanitarianism". It is a well-intentioned book, but ultimately pretty superficial.

Talking about Adam Schaff brings me to my next question, which is about Wilhelm Reich. In your book, you examine the works of numerous scholars who tend towards the psychical dimension of the individual in the individual/society duality; you make great use of their work in keeping with your own focal points and perspectives. Reich was regarded as one of the most important figures in the field during his period as a Marxist from 1927 to 1936. He produced a number of important works, from the personality mechanisms behind fascist recruitment to efforts to develop a new definition of class consciousness. He also made some important criticisms of Freud's disregard for social factors. You do not mention Reich at any point in your book, including the literature sections. Can you tell us the reason?

I think Reich is more of a lovely thinker than a great thinker. He is a warm-hearted, fascinating person, whose work is very impressive, especially in the 1930s during the Sex-Pol period. His revolutionism, his advocacy of young people's "right to make love" was so influential that it even scared organisations considered to be revolutionary or progressive into expelling him from membership. As far as I know, no other activist has managed to get himself kicked out of both the Communist Party and the Psychoanalytical Society. But even leaving aside his later years when he became absurd enough to challenge Einstein by claiming to have discovered a cosmic energy that was at the source of sexuality, Reich made no significant contribution to the theory of the individual. You can summarise his views like this: "Freud's greatest discovery is that he sees sexual drive as the most important psychical determinant. However, instead of taking this discovery to its logical conclusion and advocating sexual freedom, Freud asserted that civilization could only progress by restraining this life drive. That mistake came at a high price. The souls of the young are traumatised by the sexual dissatisfaction they are condemned to by the oppression of the state and family. Their power to embrace life head-on is destroyed, and the biggest winners in all this are capitalists and supporters of an authoritarian state. The only way to break this vicious circle and pave the way for anti-capitalist and liberatory struggle is to allow people, especially the youth, to have sexual satisfaction." How should we evaluate the viewpoint I roughly summarised? It is true that sexual oppression has both individual and social consequences. As a serious psychiatrist, Reich studied the individual consequences of this oppression, especially in his book *Character Analysis*. I ordered the book for the Faculty of Political Sciences' library at the time, and I found it a very useful read. Reich addressed the social consequences of sexual oppression in works like *The Mass Psychology of Fascism*, *The Sexual Revolution*, and *What Is Class Consciousness?* All these works are very absorbing and can almost be read in one breath. But there is not much in them other than the summary I gave above. That being the case, let us then evaluate Reich's basic thesis: is it valid that when there is sexual freedom, people will find the inner revolutionary power that is necessary to build a new life? My answer is this: it is true that sexual oppression, like all oppression, has a debilitating effect, and this effect has traumatic consequences mostly in young people. But I am not convinced that the entire mechanism of social submission can be attributed to such a factor. It would not be wrong to say that among the masses of people that states drag to war, there are many other reasons why so many "prefer to go to their death rather than revolt". Moreover, even Reich's sexual liberationism is within itself oppressive, although he himself is not aware of this. Reich, like Freud, considers only heterosexual relationships healthy. Such an approach creates a separate source of sexual oppression for those who are in the minority in terms of sexual orientation.

I think one of the most important elements of your book is that from beginning to end it is structured using a dialectical method. As far as I know, you were the first academic in Turkey to write about dialectics. Your article, "The Dialectical Method in Social Research" *was published in 1967. However, when I read your book, I saw that your understanding of dialectics has changed and progressed over time. Now I want to ask: Why dialectics? Why are you so insistent on dialectics? And in relation to that, why do you think so many Marxist social scientists in Turkey have such a poor relationship with dialectics? Of course, it is not that the dialectical method is never used in the work of Marxist academics. But it is used either partially, tacitly, or embedded in the text. At least, I am not aware of any other study constructed according to the dialectical method like your book. That is why I am asking the question.*

As you know, dialectics is a way of thinking developed by Hegel and it is hard to find anyone who has applied it more skilfully than Marx and Engels. In essence, it is based on understanding phenomena through relationality, processes, and qualitative transformation. This is its most important characteristic from a Marxist perspective. I think it would be extremely difficult to make any significant contribution to science without that kind of mindset. This applies not only to social sciences, but also to natural sciences. For example, Einstein's conception of space and time is a completely dialectical one. For Newton, space and time are each independent frameworks. But for Einstein, space is a phenomenon of "objects in space" and time is a phenomenon of "moving objects", although neither can be perceived separately. In short, the existence of everything is connected to the existence of everything; everything is moving and undergoing its own processes; a certain amount of quantitative increase leads to qualitative transformation. The idea of energy and matter transforming into each other is a striking example of dialectical thinking. However, contrary to the mistaken impression I gave in the article you mentioned, the dialectical view is not a "subject" that can be learnt in isolation simply by memorising three or four "laws" or "principles". It can only ever be learned through "application". This "application" requires, above all, a comprehensive understanding of the "science of the subject" to be studied. As Engels pointed out, if for instance the materialist conception of history is to be adopted, one must first have an extraordinarily sound knowledge of history. The same thing applies in every field. The first step towards understanding psychology, sociology, and political science through a dialectical approach is having a thorough knowledge of what has been said in these fields up until now. Basically, there is no such thing as "expertise in dialectics"; there is only "scientific, and therefore dialectical, proficiency" in any given branch of knowledge. If such a thing as "expertise in dialectics" were to exist, it would be little more than a branch of knowledge that tried to explain some abstract rules with disconnected examples. This is in itself contradictory. In other words, it is essentially impossible. How can we use disconnected examples to describe a way of thinking that by definition assumes wholeness? I disagree with the notion that dialectics "have not been successfully applied" in Turkey. That would be tantamount to saying that there is no academic science in Turkey. However, even if the scholar concerned does not consciously think "dialectically", even if they use the dialectic "tacitly, or embedded in the text", there are without doubt some excellent scholars in our country. I have a pretty good overview of the social sciences, so I could give examples from there. But it would not be right for me to do that, because once you start keeping a tally, it is all too easy to skip one name or another. It would be unfair on those who were missed out in forgetfulness or simply because their work is not as widely known. Since I was first introduced to Marxism at the age of twenty, I have always tried to use dialectics in this sense. My monograph on the *Democrat Party* and my book *An Essay on the Nature of the State* are examples of that. Now, I am trying to tackle the subject of the individual in the same way. I tried to show what the dialectical approach means in concrete when discussing the individual, both in the book's first chapter, devoted to "cognitive tools", and also in the other two chapters, where I tackle the subject by trying to demonstrate through application. It seems as if you liked my approach. I wonder how others will react.

If you do not mind, I will come back on that before moving on to my next question. I think a distinction needs to be made between "instinctive dialectics" and "intended dialectics" and that the difference is greater than people think. There could be huge differences in form and outcome between studying a subject with the view that things are directly or indirectly related as causes, conditions, or results and studying it through a dialectical construction.

For my next question, staying on the subject of dialectics, I want to ask you about a text by Marx that is at the centre of your book. I am talking about the text in the Grundrisse *titled "The Method of Political Economy" that was written as the "Introduction" to* A Contribution to the Critique of Political Economy, *but never published. You made ample use of it in your book, remarking that it was a good thing he wrote it. In my opinion, as a text on methodology, it is as valuable as it is unknown. But there is another text you mentioned, seeming to imply that it would have been a good thing if he had not written it: this is the famous "Preface" to* A Contribution to the Critique of Political Economy. *Some Marxist commentators argue that the differences between the two texts arise from the distinction between Marx's "research method" and "presentation method". I find this argument quite convincing. When I read both texts with this in mind, I can see the differences arising from the motivation behind the writing. But when I look at it more carefully, I can also see that they are not contradictory texts. What are your thoughts on the subject?*

Perhaps your objection and new question will help me to explain my point better. What made me think "if only he hadn't written it" about the "Preface" text is that it has become the main source for clumsy, mechanistic interpretations of Marxism. The reason is that it is easy to conclude from the text that the most powerful factor enabling transformation of societies is technological development. If that was truly the case, we would have to describe Turgut Özal, the viscerally anti-communist Turkish president who said that technology brought down the Berlin wall, as a Marxist! If you try to explain a rich, multi-dimensional and multi-layered viewpoint in a few phrases, you will almost inevitably end up with these kinds of misinterpretations. I wrote the article you mentioned in the last question at the same time as I was translating Politzer's book, *The Elementary Principles of Philosophy*, into Turkish. Politzer's book is indeed a "beginning", and even if it does not give you a complete grasp of dialectics, it is certainly very useful. But it is a pity that the dialectical understanding of so many generations of "Marxists" remained stuck in that place, and a heavy price has been paid for it. Because people thought a dialectical plan could be learnt abstractly and that it was feasible to construct a "planned dialectic" independent of the subject, they failed to see that a dialectical plan could only be constructed by someone with a comprehensive knowledge of the subject of study. The widespread understanding of historical materialism that arose from the "Preface" led to mistakes like this and caused extensive damage. For example, there was a noteworthy misunderstanding that dominated the Soviet Union, believing that once the revolution had taken place socialism would evolve naturally through the maximum development of the forces of production. Whereas in actual fact, it is completely contrary to Marxism and historical reality to see the forces of production as an independent factor that can have a single-handed effect. Could capitalism have achieved such a great material advance without the Renaissance and Reformation? Are petrol-rich countries with a whole host of "turnkey businesses" really developed? The truth is that for social transformation to happen, production relations together with the forces of production, and therefore the superstructure, must all change and develop together. Just because Marxists see the mode of production as the most influential social determinant, it does not mean that the mode of production, and moreover just the forces of production, can transform without society completely changing from top to bottom. To understand this properly, it is not enough to memorise dialectical rules or a few phrases from the "Preface" that summarise the materialist view of history. That is why these texts, while they might be great for memorising, ultimately cause more harm than good. And that is exactly what has happened. I also disagree that the "Introduction" shows the method of research and the "Preface" shows the method of presentation. The "Introduction" is an extremely informative text that demonstrates how both the research and presentation should be carried out in the concrete scientific field of political economy. There is no trace of the compression that is characteristic of the "Preface". I think it is worth pointing out here that Marx produced a colossal amount of work, and it would be difficult to find a scholar who is his equal. Within this huge body of work, which also includes personal drafts and scribbles that he never thought to publish, there are also other formulations that could lead to misunderstandings. For example, the passages equating feudalism with windmills, and capitalism with factories, could easily lead to a similar "forces of production deviation". We need to be really careful here. There is no place for shortcuts in science. The shiny "abstract" of an argument developed without experiencing the laborious "concrete" of research will never be any more than a deceptive glaze. The materialist view of history, and dialectics as well, can only be learnt by "doing". The subject or object of this "doing" should also be something "outside" of the view or approach. When examining a viewpoint or approach in the abstract, there is no such "doing" that can facilitate comprehension. A "tool" can only be learnt by using it on an object other than itself. Although it is of course of great benefit to read a manual beforehand. This is what you refer to as "planned dialectics". But the basic reality is that you learn to swim by swimming, and you learn to use tools by using them. Hopefully, I have managed to explain my thoughts more clearly this time.

The phrase, "The windmill gives you society with the feudal lord; the steam-mill society with the industrial capitalist", is frequently used by technological determinist Marxists and by those who criticise Marx for his technological determinism. In all likelihood, a great thinker like Marx had enough knowledge of history to know that technology was not independent from law, the characteristics of the political regime, culture, etc. I think that without knowing what type and level of abstraction Marx had in mind when he said or wrote that, in other words without knowing his dialectical method, both this statement and the text of the "Preface" can only be interpreted in a distorted way. However, since this is not our main subject, let us move on to the next question. I hope we will get the opportunity to have a separate conversation about Marx and Marxism.

The most important aspect of your book is the "feeltype theory". How did you arrive at the theory? How did you develop the idea that the psyche has six feeltypes in the same way that the body has five senses? I am very curious about the journey you took to build this theory. How do you see it?

Like every new step, it was preceded by a sense of something missing. Everybody says that humans have up until now lived in many different cultures and that there will be more cultures to come. So, on what grounds are these various cultures always said to be "human" cultures? This was the missing piece for me. The answer to this cannot be based on the physical body shared by all humankind. It would be wrong to say, "Whatever our mindset, we all have the same type of body, and that's enough to define the commonality of the human species". For that statement to be true, we would have to believe that the human species is governed by instincts imprinted onto the body just as in species of animals. However, this belief has no validity. Human beings are creatures that are governed, not by their instincts, but by themselves through behaviours defined by complying with or resisting the impositions of the cultural framework they exist in. As Marx wrote in *The Poverty of Philosophy*, and I quote from memory here: humans are "both the authors and the actors of their own drama". If these plays are so different to each other that they cannot even be compared, how can we talk of a "human species" which must necessarily, like all other species, have some common characteristics? What does dialectics tell us? It tells us: "Without similarity there can be no difference." So, for example, it would be ridiculous to try comparing a whale with a light bulb. A meaningful comparison can only be made between objects with some common aspects. That being the case, our question is this: what common points can we find in humans in order to compare the different cultures that are found everywhere that humans exist. The first answer is the body. Indeed, our body is generic. However, we do not just perceive culture with our bodies, but above all with our feelings-thoughts. The name of this feeling-thought structure found in every individual is the psyche. Therefore, the psyche is where we should seek our answer. What we call the psyche is really the individual form of cultures. Its content cannot be identical. If it was, there would be no history nor anything else. Humanity would always be a species of beings living in the same framework, and this framework could not be a culture or a "pattern of meaning"; in fact, humanity would be an animal species with a fixed nature. So, where can we find the mutual aspects of our psyches? The feeltype theory is the answer to that question. The generic commonality of all people, past and present, lies in "the possibilities of psychical sensibility". Just as the five senses of our body allow each of us to experience a limitless range of experiences while remaining within five basic dimensions, our psychical sensibilities allow each of us to experience a limitless variety of feelings-thoughts while staying within six basic dimensions. Having only five senses allows each of us to see, smell, hear, taste, and touch an infinite variety of objects. However, it blocks the door to any experiences outside of these basic dimensions. The same is true of our six feeltypes, which give us the opportunity to experience feelings-thoughts solely in the dimensions of Like/Dislike, Trust/Distrust, Approval/Disapproval, Joy/Sadness, Pleasure/Pain, Comprehension/ Confusion. The feelings-thoughts that can occur on all these dimensions have an almost infinite diversity, which varies from culture to culture. Just as the same sense of hearing can perceive very different sounds, a vast assortment of feelings-thoughts from different cultures can be perceived with the sensibility provided by one or other of the basic feeltypes shared by everyone. We still identify this vast array of different sounds as "sounds", and similarly, we can also identify a variety of different feelings-thoughts by saying, "This particular feeling-thought relates to our Like/Dislike sensibility". Without these common physical and psychical sensibilities, I think it is unlikely that we would be able to talk about "humans as a species". Let me also add that feeling-thought sensibilities are not the only issue that can be examined when investigating the psychical commonality of the human species. For example, when we consider the body, we can see that humans do not just interact with their environment through the senses, and that the kinetic possibilities and limitations of the body are also features common to all humans; the same can be said of our common "psychical behaviour capabilities" and their structural limitations. I was only able to look at the subject of sensibility.

In concrete, I reached these six axes of sensitivity by using the "trait" data developed by Gordon Allport and then others from the 1920s onwards. The idea behind these studies was to firstly compile all the words that describe "mental states" in one language, in our example, in English. Allport himself identified around four thousand words of this kind. Then these were grouped according to how closely they relate to each other. With input from various scholars, step by step the number of groups describing different basic mental characteristics was whittled down.

Eventually, they ended up with what they called “The Big Five”. According to this viewpoint, it is possible to describe any individual character by using various questions to determine the person’s traits on five different dimensions. These five dimensions are: openness, conscientiousness, extroversion, agreeableness, and neuroticism. However, I could not just take this result and use it as is, because it combines very different characteristics of personalities. On the one hand, each individual has an innate “temperament”. The types of temperament have been defined since the times of Ancient Greece, when they were described in correlation with the four basic elements of Classical Antiquity. Humans were thought to be either lively like air, introverted like earth, changeable like fire, or indifferent like water. These are traits that are passed on through heredity, just like being blonde or a brunette. However, I was only looking for characteristics that mediate the individual experience of culture, and whose content and experiential form originated from culture. That was always in my mind while researching the common sensibility axes of the psyche. That is how I reached the six axes of sensitivity, the six “feelypes”. Let me finish by saying that since the concrete contents of these emotions are different in every culture, it is inevitable that the feelings-thoughts belonging to them in that culture are unique to that culture and have different names in other cultures. Obviously, I named them in Turkish. When researching the content of these in other cultures, it should be taken into account that the related feelings-thoughts will have names specific to that culture, and as such it cannot be expected that they will be a completely faithful translation of the Turkish.

Your discovery of feeltypes is, in my opinion, ground breaking on a global scale, and it provides a pioneering guide for future studies on the subject. One of the theoretical aspects I found most interesting was that the examples you give for each feeltype are based on the relationship between one person and another. Why did you choose that type of example? I mean, is it not also valid to Like/Dislike, Trust/Distrust in our relationships with other living things or objects?

I wish all your questions were that easy. The answer is obvious: humans are the source of humans. For any person to be human they must, without exception, be in a relationship with another human being and this is an absolute principle. If there was an isolated human child in one corner of the earth, with ample food and shelter and taken care of by other animals, they would never become humanised. Contact with another human is the most fundamental condition of existence for us. Therefore, the primary source of our psyche-soul is always other people. It is the only way that feeltypes can be "stirred into processing". Once they are formed through the face-to-face relationships shaped by one culture or another, they also mediate feelings and thoughts about other objects. However, the first step in their development must always be relationships with other people. There is no place for human development in a world of "relationships" limited to animals or things. It is not for nothing that humans are defined as "the only social creatures". Society is our soul, and this holds for every one of us. We can only acquire our ability to love or fear animals and things through our face-to-face relationships with humans. That is why in all my examples of feelings, humans are the object. In all circumstances, non-human objects are always "derivative objects". Only humankind is a "substantive object".

In one part of your book you give an outstanding critique of Freud. You acknowledge his greatness, while pointing out the inaccuracies in some of his theories. At one point in your discussion, you write "the joy experienced from all types of pleasure is transmitted through the agency of the sexual drive". That leads me to a question. The sexual drive lessens and disappears after a certain age, it is not for life. But I do not think the same is true for other pleasures. How does biology make us experience these other pleasures after our sexual urges have dissipated?

If you like, we can work through the answer to your question together. One of Freud's greatest achievements was to make the human psyche the object of scientific research. It had long been known that we experience a mixture of feelings and thoughts. Freud showed that this complex reality has its own structure and that, contrary to widespread belief, most of it is unconscious. While researching the mechanism of this structure, Freud developed the view that people are generally moved by two great drives. One directs us towards life and the other towards death. Freud was a physician influenced by the French mechanistic materialist tradition, and while looking for a material basis for these drives, he concluded that the most powerful drive impelling us towards life is the sexual drive. What explains our tendency towards death, and therefore also to aggression, is the longing of the living matter that we are made of to transform into inanimate matter, almost as if it gets "tired of motion". Freud's "death instinct" argument was not widely accepted, and the pretty much indisputable power of sexuality met with more of an uncertain reception. Most scholars disagreed with the idea that the pursuit of sexual satisfaction was the most important emotion governing human behaviour. I think that objection is justified. As humans are the only social creatures, it is wrong to seek the source of their traits in the body. Human needs are determined by objects of gratification. And these objects are always socio-cultural. However, it is also important to acknowledge that human energy lives in the body. As humans are not an animal species, this energy is always mobilised through the medium of the psyche as a cluster of feelings-thoughts. When they exceed a certain intensity-concentration, these psychical motions take the form of excitements-enthusiasms. Thanks to the pleasure-pain feeltype, it is possible to experience these excitements-enthusiasms on every one of the six feeltype axes. However, the pleasure side of the pleasure-pain feeltype particularly uses the sexual elements of the body. But sexual pleasure is not only felt when stimulated by the passion of love, it can also be felt for instance when making a great scientific discovery or when a great achievement is applauded by an audience. Based on these observations, I think that pleasures originating from countless different sources can take on a sexual form when they reach a certain intensity. It may be the same observation that misled Freud. While it is true that, like all our life functions, sexual drive also weakens with age, when the feeling of an excitement-enthusiasm crosses a certain quantitative threshold, it is not realistic to say, "at this age, such an emotion can't possibly carry any sexual stimulation". If human affectivities originated from the body, from different "instincts" that can be clearly distinguished, it might be possible to say that this or that instinct disappears with age. But that is not the case with humans, and hence my answer is somewhat hesitant. Basically, immense excitements-enthusiasms can be experienced on all feeltype axes after a certain quantitative increase. Once a certain threshold has been crossed, the pleasure axis can act as a conduit of pleasure to all other axes, and at the same time it can stimulate the body's characteristic of being able to experience sexual pleasure.

There is a very lively discussion in the part of the book devoted to culture, and your analysis here requires us to rethink a lot of things. In this section, you say that the individual must firstly emerge before they can be determined by their actions. One of your main arguments is that class position itself does not determine the individual, because the individual's personality is largely determined before that through family, education, culture, and ideology. Does class position really play such a small role? When individuals start to assume their class roles, they may see the world either in terms of stocks and shares, currency parities, interest, markets and profits, or in terms of production lines, wages, house rent, and utility bills. Is it not feasible that class membership, or different positioning based on employment, has a greater impact on personality formation?

I really like that question, because it makes me realise that I have not quite explained my point. In no part of the book did I claim that class position did not determine the individual. All the face-to-face relations that determine the individual before they reach working age – family, school, friends, the dominant culture, the dominant ideology – none of them are neutral! They are all determined by class. But the influence of class here is always manifested through a number of "mediations". An individual is in a position to directly assume their class role when they become involved in the relations of production. But the personality is largely already formed by the time this age is reached. Whether the person is to be a boss or a worker is directly determined by class. But whether, as a boss or worker, they are kind or cruel is an outcome determined long before they enter the relations of production, by face-to-face relations, common emotions, cultural tools, and ideological constraints. Class position determines the individual's place in the relations of production, but the details of their psychology are a result of numerous influences starting, even before birth, in the mother's womb. Even after the individual has taken their place in the relations of production, the psyche does not reflect their class in a mechanical way. If that was the case, workers would always vote for labour parties and the exploiters of society, who are always in a minority, would never be able to win elections. The truth is, as I said in the book, that class is generally a passive determinant. For it to become an active determinant, the class struggle must be blatantly obvious, or in the words of Marx, class must not just be class objectively, but it must have become a "class for itself". In summary, class position is the main determinant in social relations. But when it comes to the formation of the psyche, class position as a general rule is manifested through the mediums of family-friends-school-popular attitudes and practices, etc. As a result of these influences, the individual acquires the main features of their personality long before they reach working age. Of course, a child might gain class consciousness at a very young age if their family, cultural milieu, etc., is very class conscious. But, again as Marx said, because the dominant ideas of any society are normally those of the ruling class, gaining class consciousness at an early age is a fairly rare occurrence. A lot of the time, individuals reach working age with a motley assortment of thoughts and feelings, only becoming more open to class influences when they take their class-determined place in the relations of production. However, even there, class generally remains as a passive determinant. Its transition to an active determinant depends on the heightening of the class struggle. Once it has become active, it becomes the most influential determinant in the political struggle; in that case, what we are talking about is a revolution being on the agenda.

I really appreciate that answer, because it neatly connects to my next question. How does a "dominant culture" form, or how does the culture of a class become dominant? And, in relation to that, I would like you to explain a little about how the dominant culture is absorbed by individuals and through which mediations?

As I wrote in the book, culture is one of the most overladen terms used in the social sciences. There is nothing that it cannot incorporate. Before answering your question, I think it might be useful to define culture as the entirety of thoughts and ideas that enable us to understand and evaluate all societal behaviours including our own. All the feeling-thoughts experienced on the six feeltype axes originate from culture. Let us take the example of the Like/Dislike feeltype in today's culture. According to the official, dominant, and most widespread feeling-thought, the greatest affection is maternal affection and children are the most valuable thing. Because of this, the family is "the backbone of society". Whereas, if we cast a critical eye at reality, we can see that hidden under this cultural mantle there is a horrific level of female oppression, from honour killings to domestic violence against women. We know that it is within the family framework that most paedophilia is inflicted on these children we consider to be "sacred" and "innocent". We can see that glorification of the family is the most powerful cultural-ideological tool for protecting and continuing the established order in all the reactionary, conservative, oppressive regimes of today, especially fascism. But how many people can see this? In the view of the vast majority of people, and therefore of the masses who are completely oppressed by the established order, this reality is just the ranting of a few rabble-rousers. This conditioning applies to every other aspect of the Like-Dislike feeltype, just as it does to every aspect of the other five feeltypes. In any society, the ideologies that reflect the world-view of the dominant classes and forces always determine what is right, what is good, what can be trusted, and what is respectable. These are repeated incessantly by the paid and unpaid spokespeople of the ruling powers, especially the media, but also by religious leaders, teachers, and opinion leaders in every corner of society. And this is how the ideas become imprinted onto the mind. The conditioning begins even when the child is still in the womb and, from the moment the child opens its eyes onto the world, this process of determination within the framework of the child's mutual interaction with society continues incessantly. From swaddling customs to lullabies, new members of every society are incessantly moulded by the influence of the dominant culture-ideology. From everyday family life to friends, school, TV, etc., this unrelenting influence continuously shapes the young individual. Behind this mass of dominant culture-ideology lies the ruling classes, social strata and the multiple institutions of the state: in short, all the social forces who profit from the established order. It is this situation that enables what Gramsci called "hegemony". Exploitation based purely on oppression cannot be indefinitely maintained. The compliance of the exploited is always required to some degree. This cultural-ideological conditioning, which starts even when the child is still in the womb and permeates all social relations, is the most powerful tool for enabling the dominant powers to secure the acquiescence of those they exploit. At this juncture, let me also mention that, for us in Turkey, religion is one of the foremost ideological weapons in getting society to adopt the interests of the dominant powers. Therefore, contrary to the beliefs of some leftists, in the Turkish context secularism is a critically important battleground for enabling the masses to see their own situation. It goes far beyond the narrow interests of the Kemalist bourgeoisie.

In your work, the theory of the individual is discussed around the relationalities of the two axes, "culture-individual", "society-individual". Could these axes not have been set as "culture-class-individual", "society-class-individual"? I want to ask this question despite the emphasis you give to social classes both in your book and in this interview.

It seems to me that behind this question lies the presumption that if a study is to be Marxist, it must pass through class mediations. If that is the case, we should firstly evaluate this presumption. It is true that in class societies, class is the main parameter determining the positions of individuals. But exposure of the determinative nature of class is a characteristic of capitalist society. The fact that class struggle was also the main determinant in previous class societies was only perceived by humans after the development of capitalism. Furthermore, even under the conditions of capitalism, this situation is only evident to those who are consciously part of the class struggle. For the vast majority of people, the reality is still blurred, and the reason is this: class is the ultimate determinant, but the determination is manifested only by passing through numerous different mediations. Looked at from this angle, I think the ordering of "culture-class-individual" or "society-class-individual" in the question may be misleading as it perhaps fails to stress sufficiently that the mediations arise from the class structure. It is perhaps better to say "class-culture-individual" or "class-society-individual"; the broadest and most powerful determinant of society is the mode of production, which in class societies manifests itself in the form of class struggle across society. The class struggle leaves its mark on all relations that make up society, all elements of culture. But in every case, this "mark" is wrapped up in a form suited to the "nature of the reality" in question. For example, in philosophy, the class struggle takes the form of a conflict between idealism and materialism. The individual's participation in society (and transformation of the society in relation to the scale of their participation) is always in the form of individual-to-individual relationships. Some of these relationships have to be face-to-face, otherwise the human juvenile could not be humanised. And the kind of relationships the society will consist of, what elements the culture will be founded on, are always determined by the mode of production. Here, however, I would like to remind you of a fact that is often overlooked, despite Marx's emphasis on it: the determinations of the mode of production (and therefore of the class struggle) always take the form of the "re-moulding" of "previously established" social relations and cultural elements. History does not "break off" just because a new mode of production has become dominant; it is constantly "re-forming". This re-formation applies to all the various roles and structures that frame and involve individuals. The individual develops through absorbing the cultural elements that dominate their environment and through face-to-face relations, which are determined by production relations on a par with the level reached by the forces of production. They internalise the effects of the mode of production and the class struggle through all these social relations and cultural elements, which have been shaped by their own unique contributing factors. Therefore, direct determination by class only occurs in extraordinary situations when the class struggle is out in the open. Apart from this, class reality, just like the level of development of the forces of production, can only determine the person by "shaping their environment". It is impossible to ascertain how this determination happens by simply relying on a general theory. Since it will always be a question of a specific class society shaping its individuals in a certain place and time, and in turn being reshaped by them, the research should always be carried out with theoretical tools whose abstractness-concreteness is appropriate to the required level of analysis. If we were to say, "In class societies, the individual is determined in such and such a way", we would remain stuck at a level of abstractness that is not of much use. But if we say, "Class determination in capitalist societies happens in such and such a way", we have achieved a level of abstraction that will be of more use. Nevertheless, that still does not lead us to a particularly valuable scientific conclusion. But if we set out to discover how individual psyches are formed in a particular class society in a particular age, we can then develop a meaningful research strategy. This would be to investigate how the individuals in various classes objectively emerge and develop in the kind of face-to-face and indirect networks of relationships necessitated by their own environments and also how their psyche forms in this mutual interaction; this must be done by taking into account the phase of history in which the relevant society exists (its position in the world), the level reached by the forces of production, the character of production relations and their class structures, while at the same time looking at how these basic data re-mould the social institutions, structures, roles, and cultural elements accumulated by the historical development specific to that society. In summary, if we want to understand the individual and society, we can try to clarify the third stage of the individual-society dialectical identicality of a certain society by using the theoretical tools that the second stage of the dialectical identicality between these two elements provides us. In other words, by asking which societal mediations the individual goes through and which individual mediations society goes through in that concrete society. Marxism, with its emphasis on class determination, can really help with this. However, in order to utilise this help, we must find out which concrete form class determination takes for each social relation (institution, structure, role) and for each cultural element (feeling-thought structures that give meaning to contexts such as family, friends, school, business relations, entertainment, politics, etc.). What will allow us to achieve this goal is not simply a general acquaintance with Marxism, but a thorough knowledge of the *specific sciences* of the individual psyche and of the various "social relations and cultural elements" that furnish the sensibilities of our basic feeltypes.

I would like to ask the last of my questions about the book's content with a question that will give us the opportunity to make a general evaluation. If the goal of Marxism, in its broadest sense, is to change the world, and if its subject is the working class, why is the theory of the individual important for both Marxism and the working class?

I want to touch upon two points. One is concerned with the agent of change and the other with the new order to be established. The agent of change is the working class. This class continually undergoes qualitative change due to the relentless development created by capitalist forces of production, described way back in 1848 in the unforgettable lines of the *Manifesto*. With an almost unfathomable foresight, in the collection of notes called the *Grundrisse*, written in 1857, Marx writes that in order to cope with the developments driven by this unbridled expansion of the forces of production, the labour force will inevitably have to use, besides nature, the knowledge created by the sciences up until that point for the act of production. And in this way, the "social individual" who will be the user of these highly developed forces of production will emerge. In such conditions, putting the labour force to work without remuneration for part of the working day, i.e. the form of exploitation based on surplus value, will be insufficient for the level of progress in the forces of production. In other words, the capitalist mode of production will be unable to cope with the progress it has itself created. This was Marx's prediction. I think, a century and a half later, his prediction has been shown to be correct. Today's advancements now necessitate that those engaged in the most advanced fields of production, especially IT, are highly skilled. These days, it is not a working class dressed in blue overalls that is dominant in the most advanced fields of production, but an entirely new working class in which mental and manual dexterity are completely fused together. The Marxist proletariat is now this class. Capitalism is becoming increasingly more fascist because it cannot accommodate this class. This rotten, capitalist order, which in its functionless condition can only bring about more oppression and destruction, must be dismantled as soon as possible for the salvation of the world. Either there will be a revolution, or we will rot away. The revolution can only happen through the leadership of this new proletariat. Now I am coming to the nature of the new order to be established. We know that according to the expectations of Marx, the new order will be communism. In his *Critique of the Gotha Programme*, written in 1875, Marx believed that this communism would be established through a continuous process, following a preparatory stage. The aim of this preparatory stage, which will inevitably carry the marks of the collapsed society, is to eradicate the traces of the old society as soon as possible and enable the development, in the forces of production – especially in the labour force – and in the production relations, that will make communism possible. In my opinion, this stage has been completed in the most advanced capitalist countries. In these countries, there is no longer any need for the dictatorship of the proletariat. On the contrary, it would be a backward step to attempt such a thing there. The relations of production have reached such a maturity that they can be transformed in the blink of an eye. There will be a direct progression to a communist order. This means the reconciliation of the hostile parties of the First International – the communists and anarchists – and the rapid implementation of radical and widespread changes like: the immediate abolition of all classes including the proletariat; the immediate replacement of the state institution, which is entirely based on oppression, with autonomous structures that everyone can directly contribute to; the disbanding of armies and the elimination of borders. Contrary to the general expectation that we must look to previous examples to see what will come next, every revolution, like every war, always takes on a very different form to the previous one. Therefore, we cannot look to the past to understand how the future revolution will be. I think it will come about through the adoption of a general stance, without the need for a lot of violence. Let us say, for instance, that one day millions of workers communicating via social media decide that in the morning they will stay in bed rather than go to work. They respond with laughter to the bluster of the ruling class. They hack the command mechanisms of the information networks used by the forces of oppression, in one swoop annihilating their power. This type of communism is a system like the one described by Marx, where "the free development of each is the condition for the free development of all". The Marxism of this kind of system will no longer be focused on analysis of class, but on analysis of the new "social individual". It is this fundamental development that has brought the theory of the individual to the top of today's Marxist agenda.

My last question is about the book's format. There is one characteristic of this book that we do not see in any of your others. You have included a section called "literature" at the end of the second and third chapters. In these sections, you critically evaluate the major thinkers who contributed to the theory of mind and culture. Why did you feel the need for this innovative format?

The format of every work is determined by its content. The reason why I included literature subsections at the end of the second and third chapters can be found in their contents. As I wrote in the book, this is such a huge subject that to look at it properly we need to refer to some of the countless sources outside of the Marxist canon. After writing that, it would have been irresponsible not to examine the works, from this vast body of literature, that contributed most to my own approach. Rather than leaving the reader swimming in a sea of ill-defined sources, I decided to write two separate subsections to evaluate the literature most relevant to my approach.

References

Boratav, K. (2004). *Tarımsal Yapılar ve Kapitalizm [Agrarian Structures and Capitalism]*. İmge.

Boratav, K. (2005). *1980'li Yıllarda Türkiye'de Sosyal Sınıflar ve Bölüşüm [Social Classes and Distribution in Turkey in the 1980's]*. İmge.

Engels, F. (1970). Letter to C. Schmidt, 27 October 1890. In *Karl Marx and Frederick Engels Selected Works in One Volume* (pp. 684–689). Lawrence and Wishart.

Engels, F. (1970). Letter to W. Borgius, 25 January 1894. In *Karl Marx and Frederick Engels Selected Works in One Volume* (pp. 693–696). Lawrence and Wishart.

Hall, C. S., & Lindzey, G. (1963). *Theories of Personality*. John Wiley and Sons, Inc.

http://www.marxists.org/archive/vygotsky/works/1931/higher-mental-functions.Htm. Accessed 31 Jan 2013.

Parsons, T. (1951). *The Social System*. The Free Press.

Index[1]

[1] Note: Page numbers followed by ‘n’ refer to notes.

C. Eroğul, *Marxism and the Individual*,
https://doi.org/10.1007/978-3-030-83662-7

Maslow, Abraham, 88, 89, 120, 121, 127n10

The manufacturer's authorised representative in the EU is Springer Nature Customer Service Centre GmbH, Europaplatz 3, 69115 Heidelberg, Germany. If you have any concerns regarding our products, please contact ProductSafety@springernature.com

Printed and bound by CPI Group (UK) Ltd, Croydon, CR0 4YY
30/09/2026
02225459-0004